Genetic Engineering
and Biotechnology

Genetic Engineering and Biotechnology

Concepts, Methods and Agronomic Applications

Yves Tourte
Honorary Professor, University of Poitiers
Poitiers, France

with the collaboration of
Catherine Tourte
PhD, Professor, BTS Anabiotec

Carole Moreau-Vauzelle
Illustrator

CRC Press
Taylor & Francis Group
Boca Raton London New York

CRC Press is an imprint of the
Taylor & Francis Group, an **informa** business
A SCIENCE PUBLISHERS BOOK

First published 2005 by Science Publishers, Inc.

Published 2018 by CRC Press
Taylor & Francis Group
6000 Broken Sound Parkway NW, Suite 300
Boca Raton, FL 33487-2742

© 2005, Copyright reserved
CRC Press is an imprint of Taylor & Francis Group, an Informa business

No claim to original U.S. Government works

ISBN 13: 978-1-57808-356-5 (pbk)

Visit the Taylor & Francis Web site at
http://www.taylorandfrancis.com

and the CRC Press Web site at
http://www.crcpress.com

Library of Congress Cataloging-in-Publication Data

Tourte, Yves.
  Genetic engineering and biotechnology : concepts, methods, and agronomic applications / Yves Tourte, with the collaboration of Catherine Tourte : Carole Moreau-Vauzelle, illustrator.
    p. cm.
Includes bibliographical references and index.
ISBN 1-57808-356-7
    1. Plant biotechnology. 2 Plant genetic engineering. I. Tourte, Catherine, II. Title.

SB106.B56T684 2004
631.5'233--dc22

                                                    2004052487

Translation of: ***Génie Génétique et Biotechnologies,*** *Concepts, méthods et applications agronomiques,* Dunod, Paris. 2002. Updated by the author for the English edition in 2003-2004.
French edition: © Dunod, Paris, 2002

Preface

The precise knowledge, control, and development of tools that will give us complete mastery of the information contained in the genome of cells seems to be the last of the grand quests of the human spirit. This advance of the biological sciences involves immense possibilities and numerous applications, but also new and important responsibilities. Biotechnology and genetic engineering are full of promise for food self-sufficiency and human health, according to some, and a formidable power, uncontrollable and dangerous, according to others. In any case, these new approaches to the living world cannot be ignored by any of us, especially those who study a biological formation of a higher order. Most often studied at the primary level, biotechnology, particularly genetic engineering, its most interesting field, must be more particularly studied at advanced levels in its fundamental aspects of biological concepts as well as its applications, which are beginning to affect our daily lives.

At first, these new technologies involved the most simple organisms, which often had small genomes, such as bacteria and yeasts. It was only later that the stable transgenesis of eukaryotic cells could be obtained and, curiously, it was in the plants that it was developed most rapidly. Transgenic crops are today numerous and are already a familiar part of our daily diet!

In its programme and its contents, this book is addressed primarily to students of biology (bachelor's and master's degrees in biochemistry, cellular biology, physiology, and biology of organisms) as well as to students of the natural sciences who are preparing for competitive entrance examinations, and teachers' examinations in biochemistry and biology. With its numerous references to agronomic applications, it is useful to students in preparatory classes for the major agronomic schools as well as to students of shorter and specialized university courses ("Anabiotech" vocational training in biochemistry and plant production, technical institutes of applied biology). Two-year diploma students who wish to pursue their studies further can also benefit from this book.

The authors felt it would be helpful to devote the first part of the book to a recapping and integration of the essentials of plant biology, agronomy, and molecular biology so as to be able to address more easily the techniques, concepts, and stakes of plant biotechnology in the various sectors in which it exerts its influence, including agro-food, health, environment, and energy.

The second part contains a discussion of the principles and practice of techniques of everything concerning *in vitro* procedures, identification, and direct and indirect transfer of genes and the control of the functioning of transgenes.

Agronomic, bioindustrial, or environmental applications that are already practised are discussed in the third part of the book. Some considerations that are deliberately prospective are raised to identify sectors in which need, or sometimes simple curiosity, will drive active expansion of these technologies in the short and medium term.

Bioethics, a moral, philosophical, and political discipline, imposed by advances in genetic engineering and the extraordinary development of these possible applications, is treated in the fourth part.

The authors wish to thank all those who, in various ways, directly and indirectly, participated in the genesis and realization of this project. They are particularly grateful to Carole Moreau-Vauzelle and Michel Bordonneau, their partners and collaborators, who invested their time and energy, the first by her contribution to the illustration of this work and the second by his critical reading of the manuscript.

Contents

Introduction

Over the past few decades, the biological literature has been enriched with a large number of new terms that serve as a support and a vector for a major field in the biological sciences—biotechnology. It is certainly a domain that is still rapidly growing but it is already so far developed that everyone knows about it. This new discipline is the fruit of accumulated results of fundamental research carried out in all sectors of the life sciences as well as applied research that have a direct impact on our everyday lives, in the fields of food, medicine, and preservation of the environment.

Biological engineering and biotechnology can be defined as the control of "living things" or the products elaborated by them *by means of technologies using living organisms or their components*. In this respect, bread, wine, and vinegar, which directly result from products elaborated by wheat or grape—flour or juice—fermented by yeast or bacteria, are perfect examples of such technology. From this we can conclude that from very ancient times people have used biotechnology without knowing it. In fact, biotechnology is commonly considered a recent achievement of research, fundamentally associated with the mastery of two major categories of molecules: *nucleic acids* that carry the information of the cell and *enzymes*, which are often the strongest expression of that information and closely involved in its exploitation.

The control of nucleic acids, i.e., their extraction from the cell without alteration of their catalytic properties, the knowledge of their molecular structure, and especially the comprehension of the role of each of their sequences in metabolism and in ontogenetic programmes, actually mark the emergence of "genetic engineering", which is today the most important part of biotechnology. This control owes a great deal to the discovery, in the early 1970s, of restriction enzymes (endonucleases) capable of recognizing precise sequences in the long chains that make up the *deoxyribonucleic acids* or DNA and cleave the molecule at a precise spot in these sequences.

A new era in biology began with that discovery. At its advent, only indirect control exploiting the different modalities of reproduction of living things was imaginable, particularly hybridization between genotypes that could be involved in sexual reproduction. These strategies involved the manipulation of considerable blocks of genes assembled by fusion of gametes during the processes of fertilization. It then became possible to test the gains and losses by successive backcrosses and stabilize these genotypes by conferring on them a certain degree of homozygosity by successive selfings. These techniques are generally time-consuming, difficult, and sometimes chancy and they can only be carried out in programmes running over years. Nevertheless, they are still the basis of plant breeding. In parallel with these difficult and therefore costly techniques, we can hope by means of genetic engineering to transfer a single gene into the genome of a plant in a single operation during its obligatory haploid phase, copy it in a second sequence in the same nucleus, and thus obtain the desired genetically modified plant right away.

Such control, in theory, has obviously not escaped the attention of agro-industrial development agencies. Many such agencies invested very early in these new technologies either by initiating a process complementary to their fundamental activity or by creating multifaceted service organizations that are more or less independent of the original institutions. Some of these structures emerged from public research laboratories that wished to improve the fundamental research so dear to their parent organizations, while helping to develop applications under the friendly pressure of the usual private partners of these laboratories and of the scientists who frequent these laboratories and wish to facilitate their professional advancement. This applied research appears to have been a powerful and effective support that has largely contributed to the development of biotechnology. Some people talk of an explosion of biotechnology in all sectors of human activity: food, energy, medicine, environment. It is undoubtedly the sudden eruption of these technologies in sectors that seem most relevant to our lives and our survival that is responsible for the popularity of this discipline, with its sometimes immoderate hopes and sometimes unjustified fears.

Chapter 1

Some Basic Concepts
of Biology

1.1. BIOLOGY AND PLANT PHYSIOLOGY

1.1.1 The range of living organisms and the place of plants

a) The major divisions of the living world

Life is closely linked to the chemistry of carbon and water. Sugars or glucides, essential components of living matter, have long been considered hydrates of carbon. Lipids are made up of three basic elements: carbon, hydrogen, and oxygen. Only the proteins have a fourth element—nitrogen, which plays an important role in the diversity of living organisms. All these elements also belong to the inert mineral world, solid or gaseous, and are constantly involved in exchanges between that inert world and the living world. These exchanges and matter flows, which consume a large amount of energy, constitute the major cycles of water, carbon, and nitrogen. "Life" seems to be the essential driver of such periodic cycling of elements.

The extraordinary diversity of living forms has necessitated the conception of a classification of these forms, putting into a single group those individuals that share a certain number of common characteristics. This is how, for example, the *prokaryotes*, which like bacteria do not have a nuclear compartment, are distinguished from *eukaryotes*, which carry most of their information in a territory called the *nucleus*. Eukaryotes are themselves subdivided into *plants* and *animals* according to whether or not they are equipped and able to use light energy. A preliminary approach to the classification of living organisms is thus based on three major, clearly distinguished groups. There are also intermediate groups, often constituting interesting transitions that merit a closer look.

b) Definition of a plant

Generally, a plant is defined by its capacity to use light energy, i.e., energy linked to light radiation, to combine water molecules and carbon atoms to synthesize sugars. The sugars constitute energy reserves that allow plant cells to carry out all the other syntheses necessary for their metabolism. This carbon-based *autotrophy* of plants is based on the presence of specific pigments such as chlorophylls, while animals, which do not have this capacity for synthesis, use plant sugars and are thus entirely dependent on plants. They are situated further along the food chain. This essential criterion, however, is not the only one and normally the following criteria are added, in no particular order.

• the existence of meristems, which give plants a constant and indefinite growth;

• a general totipotency of cells that preserves the capacity of *regeneration* throughout the life of the plant, i.e., the ability to recreate a new plant;

• meiosis, which results in the differentiation of spores and not gametes;

• an alternation of generations, *gametophyte* and *sporophyte*, separated by fertilization and meiosis;

• the presence of rigid framework of cellulose around each cell;

• the existence of a triple heredity—nuclear, plastidial, and mitochondrial;

• a kinetic apparatus that has some variability in its organization, for example, centrioles replaced by polar crowns;

• male gametes that do not always have the capacity for autonomous displacement;

• a fixed life.

c) Frontiers of the plant kingdom

➤ With prokaryotes

Some bacteria are also photosynthetic because of the presence of chlorophyllian pigments (bacteriochlorophylls). At this frontier of the plant kingdom are found the *Cyanobacteria* or blue bacteria, which were earlier called *blue algae*. They are bacteria because they have no nuclear compartment but are classified among the algae on the basis of the vegetative organization, which can be as evolved as that of other algal groups (nematothallus and cladothallus). These cyanobacteria are also distinguished by the differentiation of a complex photosynthetic structure, the chromoplasm, which associates concentric membranes and pigments dissolved in lipidic drops (Fig. 1.1). They are perfectly autotrophic for carbon. The genera *Nostoc* and *Oscillatoria* are representative of these organisms.

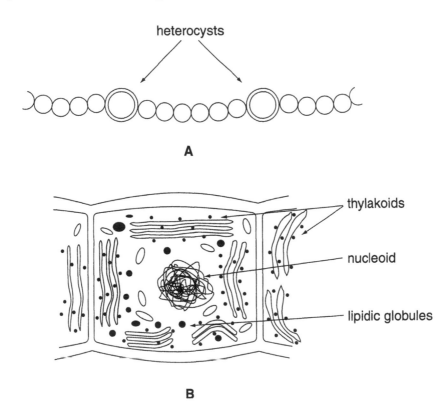

Fig. 1.1. (A) Colony of a cyanobacterium, nostoc, with its heterocysts. (B) Details of a cell with its chromatoplasm composed of thylakoids, nucleoid, and lipidic globules.

➤ With the animal kingdom

Although there is no difficulty in recognizing the kingdom to which a rabbit or a carrot belongs, the limits are fluid and some groups can pose problems. Three examples are taken up below.

The Euglena are generally considered unicellular green algae. They are autotrophic because of the presence of numerous plasts enclosing pigments associated with membranes. However, they are mobile, because they do not have the usual cell wall found in plant cells and do have a flagellar part. When cultured in conditions highly favourable to their multiplication, the population grows exponentially but the rate of nuclear division is higher than that of chloroplast division. Thus, there is a gradual diminution in the number of plasts per cell. This phenomenon recurs until the Euglena has only one chloroplast at the time of mitosis. The result is an unequal division: one of the cells inherits a chloroplast and the other does not and becomes the source of an aplastidiate cell line. If this line has a source of carbon in its culture medium, it multiplies as do protozoa in culture and thus belongs irreversibly to the animal kingdom.

Some organisms at the frontier of the two kingdoms are similar to one another, but some are photosynthetic and some are not. This is the case of many flagellates as well as some algae, the Pyrrophycophytes, and protozoa belonging to the Rhizopods. The two groups have the same type of alternation of generations. An alternation of generation, although typical of plants, characterizes the cycle of trypanosome, a protozoa that causes malaria.

Another problem is that of the Mycophytes or fungi, which play a very important role in biotechnology. Even though they are not chlorophyllian, they are generally considered very similar to plants because of their fixed life, their alternation of haploid-diploid phases, the formation of meiospores, their cell wall, or their plant-like organization, which puts them in the category of Thallophytes. However, their heterotrophy, the presence of chitin in the composition of their cell wall, the existence of centriolar parts to direct mitosis, and a mobile plasmode phase in the group Myxomycetes (also called Mycetozoa by zoologists) are arguments that can be opposed to botanists who insist on considering those organisms as plants. It is preferable, in fact, to consider the fungi an independent and original group between plants and animals.

d) The major plant groups and biotechnology

It is presently difficult to propose a classification of plants that would be accepted by the entire community of botanists. There is a constant evolution, not to mention a constant revolution, in the science of systematics. Therefore, though our position may be obsolete, it has the merit of being still accepted by most people and compatible with most studies.

In the system followed in this book, plants are divided into two major groups depending on whether they possess an axis that provides support, distribution of water and metabolites, and an erect habit. This axis (or trunk) allows us to distinguish the *Cormophytes* or higher plants from *Thallophytes* or lower plants. The following dichotomies are used to make subdivisions within these two groups.

- within the *Thallophytes*:
  - Algae,

- Fungi,
- Lichens.
• within the **Cormophytes**:
  - Cryptogams:
    * Bryophytes or Mosses and related plants,
    * Pteridophytes or Ferns and related plants,
  - Prephanerogams: Ginkgoales and Cycadales,
  - Phanerogams:
    * Gymnosperms or Conifers and related plants
    * Chlamydosperms: Gnetales and related plants,
    * Angiosperms:
       Monocotyledons or *Liliopsida*,
       Dicotyledons or *Magnolopsida*.

Only some of these groups or classes of plants are of interest in biotechnology. We will have occasion to note this as we describe the technical approaches and bioindustrial uses of these plants. We can, however, as a preliminary approach, characterize the relationships between these various groups and biotechnology and report the following general indications.

The *Algae* or Phycophytes now constitute a field of application of biotechnological research involving experiments to obtain cultured organisms, cells and calluses the commercialization of some raw materials designed for animal and human food, additives (agar, alginates, carrageenans), and products that have pharmaceutical (particularly antiviral) and cosmetic applications. Genetic engineering is still in its infancy in this group.

The *Fungi* or Mycophytes, particularly yeasts and Ascomycetes, are of particular interest in genetic engineering and biotechnology. The entire genome of bakers' yeast (*Saccharomyces cerevisiae*) has been sequenced and most of it has been decoded. We will see that these yeasts are most valuable auxiliaries in acquiring knowledge about and mastering plant genomes.

In the *Cormophytes*, the biotechnological approach is still very weak with respect to the Bryophytes (mosses and related plants) and Pteridophytes (ferns and related plants). In the Prephanerogams, the Ginkgo presently holds a great deal of interest because of the presence of certain diterpenes with significant pharmacological effects. For similar reasons, much attention has been given to Gymnosperms (conifers and related plants) and especially *Taxus* (yew), which yields taxol, an extract with known antimitotic and thus antitumoral properties. The synthesis of such products by genetic engineering is important in current research. It is probable that the equally active search for pharmaco-dynamic molecules will lead the pharmaceutical industry to look at Chlamydosperms and *Ephedra* in particular.

The Phanerogams are by far the most interesting group in terms of biotechnological approaches and, consequently, the most often cited group in this book. It includes almost all plants of agronomic and horticultural interest and is, of all living organisms after bacteria and yeasts, the group in which genetic engineering seems to have advanced most and for which economic considerations seem to have the greatest weight.

1.1.2. Plant growth and development

a) Ontogenic programmes

The structure of a plant corresponds to a set of morphological and physiological phenomena that occur in a precise order and exploit information inscribed in the genetic inheritance of each individual: this is called an *ontogenic programme*. In fact, there are two successive programmes in plants, since there is an alternation of generations: a *sporophytic ontogenic* programme for the plant strictly speaking and a *gametophytic ontogenic* programme during the appearance of reproductive processes. Each programme comprises processes of growth and development.

Growth is characterized essentially by a phase of intense *cell multiplication* followed by a phase of *cell elongation*. The first is *meresis* and the second is *auxesis*. It is essentially an increase in biological mass by true reproduction of cells.

Development corresponds to the successive appearance and *differentiation* of the organs of the plant: stems, branches, roots, leaves, and flowers. The differentiation of organs requires highly varied modes of cell and tissue differentiation.

The progress of an ontogenic programme requires an initial element or structure that can vary greatly in plants. This element is very often unicellular: a *zygote* for the sporophyte and a *spore* for the gametophyte. It can also be multicellular: fragments of tissues or organs as well as similar parent cells of gametes that could give rise directly, by parthenogenesis and true multiplication, to a new organism, without participating in a phenomenon linked to sexual reproduction.

Usually, there are two periods in the life of a plant: *growth and development of vegetative parts* and then acquisition of the capacity to reproduce sexually, which is the *building of a reproductive apparatus*. Like any living organism, the plant then ages and dies and its matter returns to the mineral world through *processes of humification* before being cycled again by the development of new organisms.

b) Growth and development of vegetative parts

The growth and development of vegetative parts generally begin with the fertilization and formation of a zygote, the modalities of which will be discussed later. Let us note, however, that there is an apparent paradox here, since there is normally an opposition between the concepts of "totipotent" and "differentiated" cells; here, on the other hand, we have the meeting of two highly differentiated cells, the gametes, which on fusion give rise to the quintessential totipotent cell, the zygote. This cell will very rapidly divide according to various modalities peculiar to each species. Nevertheless, it results in most cases in the establishment of two cells of different sizes: the larger is the origin of the embryo, and the smaller develops into an organ of very unequal size serving as a link between the embryo and the mother plant: the suspensor. The embryo cell will divide repeatedly into a small cell mass that is usually spherical, or a globular embryo, then very quickly there will appear an anterior-posterior polarity, associated with a plane of bilateral symmetry, indicating the position of all the primary organs: one or two *cotyledons*, the *radicle* or future first root, and the *stemlet* terminated by the *gemmule*, the origin of the future stem. The gemmule is in fact the virtual organizational link of almost all the organs of the plantlet (Fig. 1.2). This first development of the embryo occurs within the *seed*, a descendent of the ovule, which is itself generally protected by the *fruit*, the heir of the ovary of the flower, at least in the Angiosperms. The development could be continuous but most often it slows down which indicates a pause coinciding with the arrival of an unfavourable season: this is *dormancy*, a phenomenon in which there are various factors involved: external (temperature, light, day length) and internal (hormonal and trophic). Lifting of dormancy with a resumption of mitotic activity and appearance of new organs coincides with the return of the favourable season. Some plants need a cold period of some duration and intensity in order to resume mitotic activity. This is the characteristic *vernalization* of so-called winter plants: wheat, rye, rape, in which the lifting of dormancy necessitates the accumulation of a certain number of degrees of cold. Here also, external factors (temperature, light, humidity) combine with internal factors and the hormone abscisic acid has a particular role in reviving metabolic activities.

On the morpho-anatomical plane, growth is resumed from the *primary meristems* located in the distal tips of the axis: root meristem and shoot meristem. There are zones of major mitotic activity that ensure the elongation of the organism. There is also in this axis a region of cell elongation without division, thus with a constant number of cells, which can also be located just below the insertion points of cotyledons

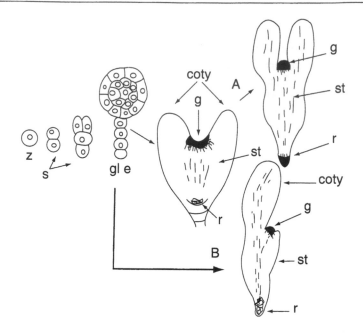

Fig. 1.2. Embryo development in Phanerogams. The zygote (z) differentiates a suspensor (s), while the larger of the first two blastomeres will result in a globular embryo (gl e), then the embryo with bilateral symmetry with cotyledons (coty), gemmule (g), rootlet (r), and very short stemlet (st). (A) An embryo of a Dicotyledon and (B) one of a Monocotyledon.

(epigeal growth, as in castor, some beans, radish, and tomatoes) or just above them (hypogeal growth as in peas, faba bean, maize, or wheat). In the former case, the cotyledons become aerial and in the latter they remain underground (Fig. 1.3).

The meristems are the origin of all plant tissues: they are called *histogenic*. They are also involved in the formation of organs, leaves, and branches. This is their *organogenic* role. The internal organization of meristems, particularly of the shoot meristem, is generally heterogeneous and two major zones can be distinguished: an axial zone in which the cells have a very low rate of division and a peripheral zone in rings within which the rate of mitosis is much higher. This is called the *initial ring*. Within this ring, with a particular periodicity, there appears a leaf *initia* that grows into a *leaf primordia* and then into young leaves (Fig. 1.4). The time that separates the appearance of two successive initia is called *plastochrone*. It is specific and characterizes the modalities of leaf insertion along the stem: alternate, opposite, or whorled leaves. The study of these modalities and mechanisms that determine leaf insertion is called *phyllotaxy*, a science that was revived

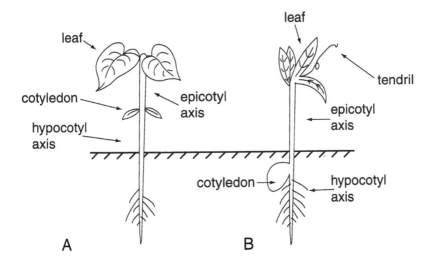

Fig. 1.3. (A) Epigeal growth of the bean. (B) Hypogeal growth of peas.

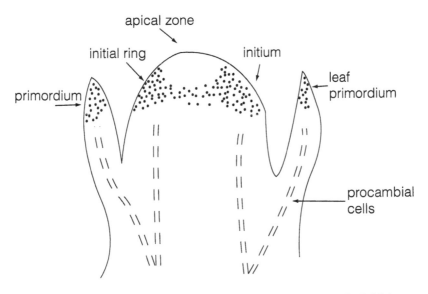

Fig. 1.4. Vertical section of a stem apex. The apical zone caps the initial ring, which gives rise to young leaves that pass successively through the stages of initia, primordia, and leaf primordia. The procambial cells begin to differentiate.

in the middle of the 20th century by the efforts of the botanist L. Plantefol and his colleagues. These meristems also give rise to all the stem tissues: the protective tissue or epidermis, the supply or reserve tissue (parenchyma, pith, aquiferous and aeriferous tissues), sap-conducting tissue (xylem for raw sap and phloem for elaborated sap), support tissue (collenchyma and sclerenchyma), and tissues that secrete numerous products of secondary metabolism.

Meristems function throughout the life of the plant but with an intensity that varies with the seasons: histogenic and organogenic activity is high in the spring and summer and slows down during autumn. It may stop completely from the first frost onwards. Buds during winter have a particular physiology that allows them to resist frost more or less effectively.

There are also structures that allow growth in thickness. These are called secondary meristems. Nearly non-existent in most herbaceous species, such as herbaceous grasses, they are particularly active in dicotyledonous woody species. They are generally present in the form of two concentric circles: the inner circle, located near the conducting apparatus, is called the *libero-ligneous layer* or *cambium* and is found to be the source of secondary conducting bundles of two types of saps as well as the reinforcement of support structures (wood fibres). The outer layer, or *suberophyllodermal layer*, is involved in the development of protective tissues and is the origin of the secondary bark or *periderm* (Fig. 1.5).

Physiologically, the functioning of meristems and the establishment of all these tissues is based on *trophic elements*, the provision of which is ensured by root absorption and photosynthesis, and *growth factors* or *hormones* that are synthesized for the most part within the growth tissues themselves.

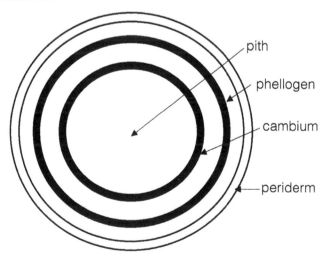

Fig. 1.5. Cross-section of an older stem showing the concentric positions of two generating layers (cambium and suberophellodermal layer).

Aquatic plants absorb water through their entire surface, while terrestrial plants absorb water and some minerals from the soil through their roots and absorbent hairs and distribute them throughout all their cells by means of xylem vessels. There is thus an ascending water current between the roots and above-ground parts, essentially the leaves; a certain quantity of this water is constantly eliminated in gaseous form (transpiration) or liquid form (guttation). This water is eliminated through specialized structures called the *stomata* and *hydatodes* and the process is actively involved in the transport of dissolved salts. Dissolved salts are absorbed in an ionized form in proportions that follow a specific equilibrium for each plant. This absorption consumes energy and requires particular transporters. It responds to the needs of the plant for elements that constitute its most complex molecules (iron, magnesium, calcium, nitrogen, etc.) and cofactors indispensable to a large number of its metabolic reactions. The ammonium ions indispensable for the synthesis of amino acids is also supplied by root absorption of products from the metabolism of organisms that can use atmospheric nitrogen, such as certain bacteria and cyanobacteria that live in the soil or in symbiosis with the roots themselves (legumes, elms, etc.). The genetic control of stages of the nitrogen cycle is one of the most interesting subjects of research for biotechnologists. This subject will be discussed again in Chapter 2.

Apart from the parasitic plants, which draw their carbon in organic form from the sap of their hosts, chlorophyllian plants get carbon from atmospheric carbon dioxide through *photosynthesis*. Without going into detail about this complex but now relatively well-known mechanism, let us recall some essential points. The linking of carbon with an association of hydrogen and oxygen occurs in the chloroplasts and consists in converting the energy from solar radiation into chemical energy. This conversion is ensured by molecular complexes, structured according to a periodic organization within the internal membranes of the chloroplast, and associating structural and enzymatic proteins to the chlorophyll molecules.

There are several types of chlorophyll, characterized by slightly different parameters of light absorption (Fig. 1.6). These molecular complexes constitute the photosystems I and II associated with ATP synthetases. The photosystems are made up of a peripheral antenna charged with collecting photons and a reaction centre within which the chemical reactions themselves occur. Photosystem II collaborates with a complex rich in manganese to destroy water molecules, recover the electrons and protons from this photolysis, and transfer their energy to molecules such as ADP and NADP. Some oxygen is thus liberated by the photochemical reaction, while the protons from the hydrogen accumulate in the closed space that constitutes the intermembrane

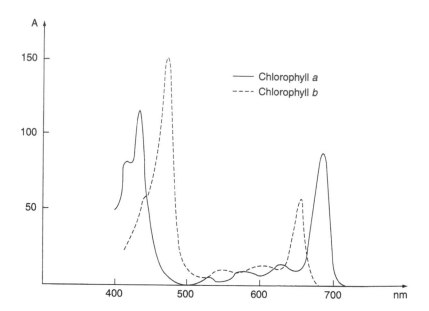

Fig. 1.6. Absorption spectra of chlorophyll *a* and *b*; The wavelengths in nm are shown on the X-axis and the light absorption (A) is shown on the Y-axis (Heller et al., 1998-99).

space of the thylakoid. A gradient of proton concentration forms between this space and the stroma of the chloroplast, creating a return flow of protons towards the stroma. It is the electrochemical potential created by this proton flow through the ATP synthetases that is recovered to realize the phosphate linkages on the energy molecules (Fig. 1.7). This first part of photosynthesis that uses light energy is called the *light* or *luminous phase*. It is followed by a phase called the *dark phase*, located in the stroma, during which the metabolites previously synthesized are consumed to fix the carbon dioxide on the sugar molecules. Thus, in most cases a C_5 sugar, ribulose 1-5 diphosphate, plays the role of CO_2 acceptor and gives a C_6 sugar that immediately divides into two C_3 molecules, of 3 phosphoglyceric acid (Fig. 1.8). This last reaction is catalysed by an enzymatic complex known as "rubisco", the formation of which associates proteins synthesized in the chloroplast itself and others coming from cytosol. The mastery of this enzymatic complex is also of interest to geneticists. There are also variants of this type of metabolism, particularly in tropical plants (C4 and CAM plants).

The products of photosynthesis elaborated by leaf parenchyma cells are then distributed to all the plant cells through the elaborated sap circulating in the phloem cells. Such transport requires energy and the presence in each plant of transport molecules specific to each sugar.

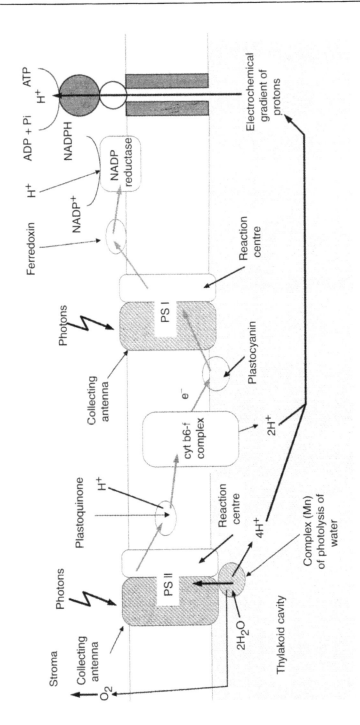

Fig. 1.7. Relative positions of photosystems I and II, electron transporters, and the synthetic ATPase complex in the thylakoid membranes

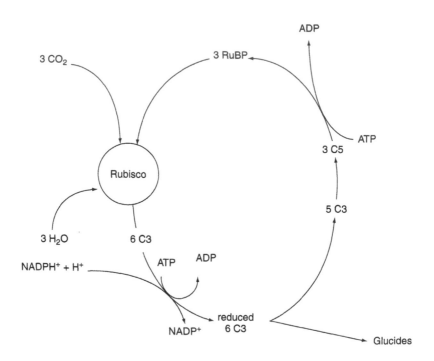

Fig. 1.8. Role of rubisco in the input of CO_2 and in the regeneration cycle of ribulose 1-5 diphosphate (RuBP); C3, 3 phosphoglyceric acid; reduced C3, 3 phosphate gylcerate; C5, ribulose 5 phosphate.

These transporters between place of production or *source organs* and the places of use or *sink organs* are also the subject of important biotechnological research.

Plant growth follows rules that fix the general harmony of the plant and consequently influence its habit. These rules are part of the genetic inheritance of each species and are expressed through chemical messengers called growth factors or hormones, although these products do not answer exactly to the definition for animal hormones. There is in fact sometimes a coincidence between source and sink organs. The products are not always secreted in the circulating liquid, which is here the sap as well as the apoplasm, represented by the walls, or they are secreted in the external environment, as with ethylene. More than the presence of molecules, it is the equilibrium between the different families of hormones that plays a dominant role and ensures the true correlations between organs during development. The first plant hormone identified was *indole-3-acetic acid* or *IAA*, which was called an *auxin*, a factor of stem elongation (Fig. 1.9).

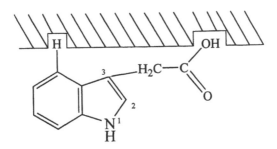

Fig. 1.9. Molecule of IAA or indole acetic acid and its modalities of fixation on its receptor

Today, the term *auxin* comprises a number of molecules of structure very similar to that of IAA, produced naturally or by synthesis and having comparable effects on growth and development. It was very early demonstrated that the movement of these hormones was polarized between the source organs, apex and intercalary meristems, and the target tissues, the stems, leaves, and flowers, or the accumulation zones, the roots. These molecules, which are derived from tryptophane, act at very low doses (of the order of 10^{-7} g ml^{-1}), are photolabile, and are easily degraded. They are involved in tropisms or directional growth, in cambial activities, and fruit development. We will see that biotechnologists rely extensively on these hormones in *in vitro* culture of cells and tissues as well as in the perfection of processes of plant regeneration. These hormones are also responsible for apical dominance, a phenomenon by which the development of buds located on the leaf wings and branches (axillary buds) is usually inhibited when the terminal bud is growing. The excision of the terminal bud leads to lifting of the inhibition, until one of the axillary buds, generally the one located closest to the terminal bud, takes over, thus re-establishing the dominance. The programmed size of trees has a precise effect on such lifting and re-establishment of hormonal correlations to modify the general habit of the plant.

The so-called "synthesis" hormones and particularly the best known, 2,4–dichlorophenoxyacetic acid or *2,4–D* (Fig. 1.10), causes unbalanced growth of plants, also called *morphosis*. At high levels, these hormones are used as selective herbicides, since all plants do not have the same sensitivity to them. Dicotyledons, generally very sensitive to them, are eliminated from fields planted with cereal crops, which, like all Poaceae or Gramineae, are less sensitive to them. Today, it is acknowledged that these hormones are more effective when their spatial structure is active over two sites of fixation, as with IAA.

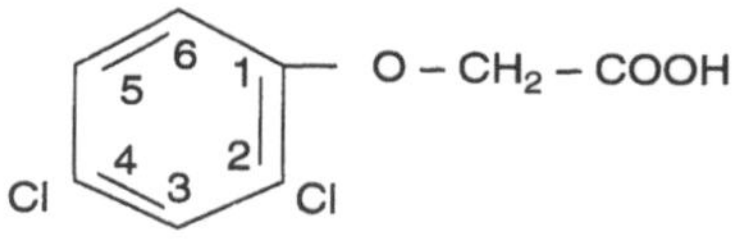

2,4-D

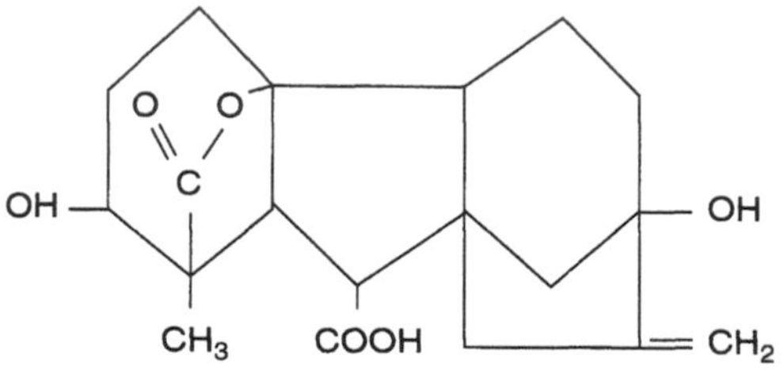

GA₃

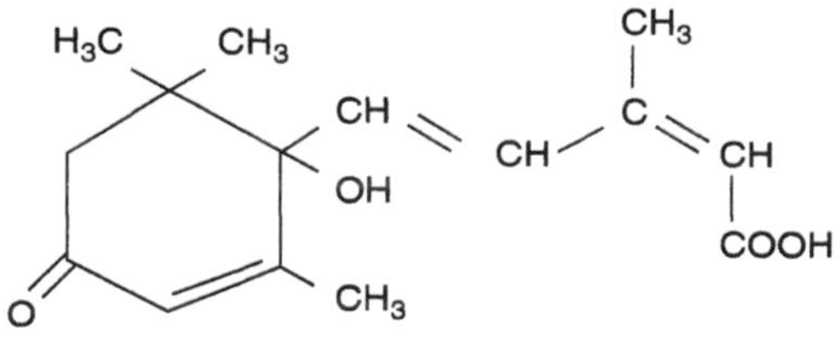

Abscisic acid (ABA)

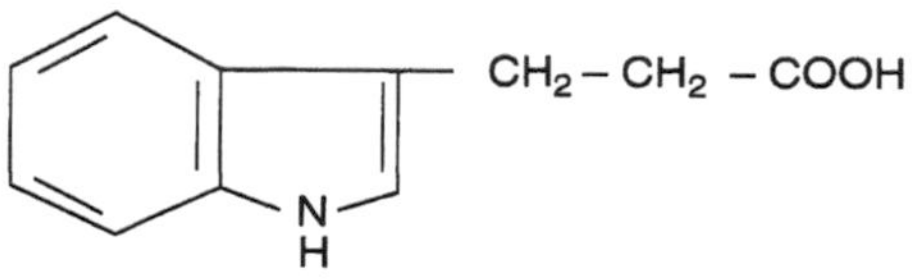

Indole-propionic acid (IPA)

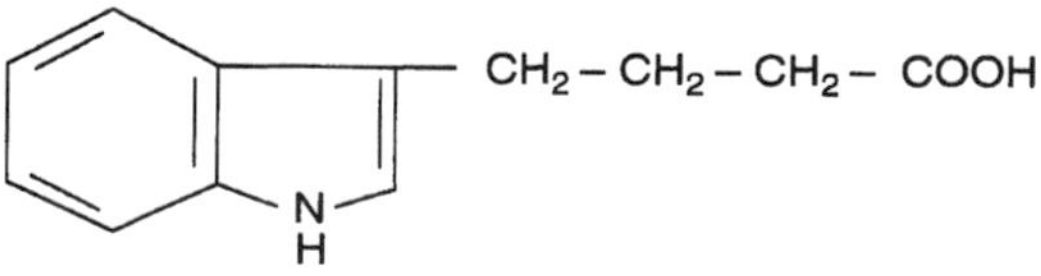

Indole-butyric acid (IBA)

Fig. 1.10. Formulas of three synthetic hormones (2,4-D, IBA, IPA), gibberellic acid (GA3), abscisic acid (ABA) (Heller et al., 1998-99).

There are other groups of growth factors from other pathways of biosynthesis. One is the *gibberellin* group, the best known example of which is extracted from a fungus, *Gibberella fujikuroi*, which is responsible for spectacular growth of internodes in rice. Other gibberellins were subsequently discovered in nearly every plant group and even in bacteria. They are denoted by numbers, GA_1, GA_2, GA_3, etc., and more than sixty have been identified. They all are derived from the terpene biosynthesis pathway, particularly the geranylgeraniol step. Gibberellins stimulate meresis and auxesis of stems and leaves and influence the processes of flowering. They are also found in the lifting of dormancy in seeds and embryos, acting as an antagonist to *abscisic acid*. They are said to play an active role in the synthesis of the enzyme α-*amylase*, which is involved in the hydrolysis of starch into assimilable sugars required for embryo development. In *in vitro* culture, the ratio of auxins to gibberellins must always be studied with attention.

Another group of growth regulators, the *cytokinins*, was identified in the mid-20th century. The first cytokinin to be studied was extracted from coconut milk, a product that is still often added to tissue culture medium, and named *kinetin* because it favours cell division or kinesis. Other cytokinins were identified and studied from a wide variety of tissues. Examples are zeatin, an extract of maize (*Zea mays*), and benzylaminopurine (BAP), obtained by synthesis, which are largely used in *in vitro* culture. They are generally formed from a purine nucleus analogous to that of adenine, which constitutes one of the nitrogenous bases of DNA and from a lateral terpenic chain derived from mevalonic acid. More than 30 cytokinins are now known. These are present in nearly all plant tissues but in greater quantities in roots, where they are synthesized, as well as in fruits and seeds. They are involved in growth in that they stimulate mitosis and in some cases also in auxesis and development through the lifting of dormancy of seeds, activation and formation of buds, and genesis of floral primordia. As with the gibberellins, it is the cytokinin-auxin ratio that in fact plays the most important role in determining the precise direction of development.

Abscisic acid or *ABA* (Fig. 1.10), which is a sesquiterpene, is also considered a plant hormone antagonistic to those mentioned earlier, particularly gibberellins, with respect to its effects on growth and development. It is named for its involvement in the abscission of leaves and fruits. It also has an effect on the dormancy of buds and seeds, as well as on the activity of stomata and, consequently, the intensity of transpiration.

Ethylene can be considered the last substance to enter the plant hormone club, even though its role in fruit maturation has been known for a long time. The banana ripening rooms installed to provide town gas regardless of the security of the neighbourhood will be recalled. As they matured, the fruits excreted large quantities of ethylene that directly influenced the maturation of other fruits nearby. It also affected leaf fall in autumn. Its synthesis is complex and has a distant precursor in methionine. Many cases are known of synergy of this product with the previously mentioned hormones. The control of ethylene production is one of the most remarkable successes of biotechnologists, as we will see later.

1.1.3. Plant reproduction

a) Modalities

In higher plants, reproduction can occur according to various modalities that, in some cases, cause a group of more or less differentiated cells to divide into various organs: stem, root, leaf, and, in other cases, some cells that develop into gametes, located in a specialized organ, the flower. The first is *somatic reproduction* and the second is *sexual reproduction*. Some particular modalities could constitute a third modality involving gametic cells or close relations of gametes, but participating only partly in the traditional modalities of sexual reproduction. This is *asexual reproduction*. The relative importance of these different modalities varies with each family, genus, or species. We will thus see successively:

• the modalities of sexual reproduction;

• asexual reproduction;

• vegetative propagation and somatic regeneration of plants.

b) Sexual reproduction of plants

In the most evolved plants, Gymnosperms and Angiosperms, sexual reproduction occurs in a particular organ, the flower. This organ often combines in a single structure the elements that give rise to male gametes or *sperm nuclei* and those that differentiate the female cells, the *oospheres*. This highly frequent bisexuality of the flower does not imply that the meeting of gametes is limited to this closed field; the plant has many mechanisms, structural as well as physiological, to ensure a certain levelling of gametic encounters and, consequently, a genetic mixing of the population to which the plant belongs.

Fundamentally, a flower, in its most classical structure, comprises four whorls of floral parts inserted at the tip of a short branch that has the appearance and structure of a stem, the *floral peduncle*. This peduncle is often swollen in the distal part, presenting a structure that looks like a club, a disc, or a chalice (Fig. 1.11).

On this swollen part, or hypanthium, is inserted the first outer whorl of sepals: generally four or five in Dicotyledons and multiples of

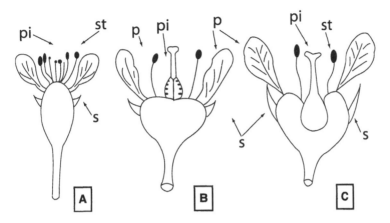

Fig. 1.11. The three basic forms of the hypanthium: (A) club-shaped or thalamus, (B) disc, and (C) chalice or conceptacle. s, sepal; p, petal; st, stamen; pi, pistil.

three in Monocotyledons. These sepals, most often chlorophyllian, belong more often to leaves. The next whorl consists of petals often numbering as many as the sepals. The third whorl contains the stamens or male organs, the swollen upper parts of which, the anthers, contain pollen sacs that are in turn connected to the peduncle by the filament. The centre of the flower, corresponding to the fourth and final whorl, is occupied by the female organs, often united, and constituting the pistil, the lower part of which is the ovary, surmounted by the style and stigma (Fig. 1.12).

The number of united carpels forming the ovary is not always closely related with the number of parts in the other whorls.

This example, which is only a general description, is one of a regular flower of type 5 with united carpels. In reality, flowers show a wide diversity of form, symmetry, and organization. They can be either regular (*actinomorphic*) with an axial symmetry, such as the flower we have described, or irregular (*zygomorphic*) with bilateral symmetry, such as the flower of the snapdragon or orchid. The flowers are isolated or grouped into inflorescences (cymes, clusters, capitula, etc.).

All these arrangements and structures constitute *floristics*, which plays a critical role in the classification or *systematics* of the plant world. The reader who is interested in a deeper study of these disciplines is referred to books on botany (e.g., *Plants* by Irene Ridge, Oxford University Press, 2002).

The mechanisms that control the birth of a flower have been the object of innumerable studies of morphology, anatomy, physiology, and genetics, studies that have sometimes ended in varied and often

A

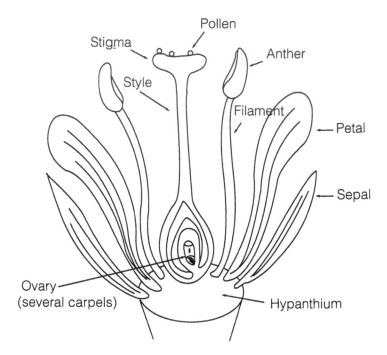

B

Fig. 1.12. (A) Diagrammatic section of a flower with a disc-shaped hypanthium with its four whorls of floral parts. (B) Floral diagram of a flower of type 5.

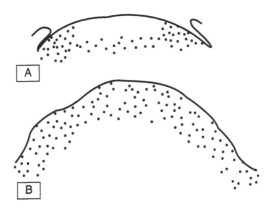

Fig. 1.13. Comparative illustrations of (A) a vegetative meristem elaborating leaves with its initial ring and (B) a meristem at the beginning of floral initiation, showing a homogenization of mitotic activities.

contradictory theories. The time is not far off in which there will be arguments in support of various "schools" on the origin of the flower. Some researchers, in the wake of the 18th century writer Goethe, conceive it as a simple "metamorphosis" of leaf organs. Others, on the basis of the studies and ideas of the botanist L. Plantefol, propose that it is an original organ resulting from histogenic and organogenic activity of a meristem in waiting unobtrusively hidden at the tip of a flowering branch.

Although the physiological and genetic mechanisms of the formation of a flower are not yet totally elucidated, they are increasingly better understood.

Anatomically, the meristem of the axis that will give rise to a flower undergoes a profound structural revolution, mostly corresponding to a loss of the zonation that characterizes the vegetative meristem and that tends gradually towards a mitotic homogenization (Fig. 1.13). This important mitotic activity leads to the building up of several initia that, depending on the positions they occupy on the meristem, are the source of various floral parts (Fig. 1.14).

Physiologically, the factors that contribute to flower set are numerous. Some are external, linked to the climatic conditions that dominate throughout a certain period: essentially rainfall, temperature, and light. Temperature and light act not only by their intensity but also and primarily by their duration and rhythmicity. This is referred to as thermoperiodism and photoperiodism. Studies in these fields have ended in a classification of plants in relation to their needs: short-day or *nyctoperiodic* plants and long-day or *hemeroperiodic* plants. Other factors are internal, closely correlated to the state of development of the

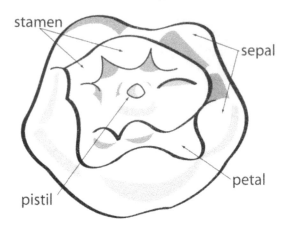

Fig. 1.14. Three-dimensional drawing of a reproductive meristem during the formation of different primordia of floral parts.

vegetative parts. Flower set corresponds, in fact, to particular quantitative and qualitative trophic needs. It can be considered that the conjunction of a set of external and internal factors, at a given moment, is responsible for *floral initiation*, which represents the very first manifestation of the floral primordium, the realization of the flower or *floral organogenesis*. Among all the molecules implicated in this realization, *phytochrome* seems to play an essential role in flower set (Fig. 1.15). Phytochrome is an important photoconvertible photoreceptor system, sensitive to red and infrared light depending on its state of excitation.

Under the action of red light (660 nm) and infrared light (730 nm), the phytochrome molecule changes reversibly from an active to an inactive form.

With respect to the genetic control and development of the flower, advances in understanding have been much more recent and spectacular. They owe a great deal to the choice of the *Arabidopsis* model. The complete gene sequence of this plant was published at the end of the year 2000 by the community of molecular biologists. The choice of this plant led to manipulation of a considerable number of individuals and, consequently, the location and selection of many floral mutants. The need to classify these mutations and to be able to interpret their causes led some researchers to propose a floral ontogenesis diagram that in some ways recalls the embryo development of animals (Fig. 1.16). It is in *Arabidopsis* that the FLT gene was discovered in 1999: it determines the transition from the vegetative to the reproductive phase. In fact, when there is a mutation, the FLT gene is

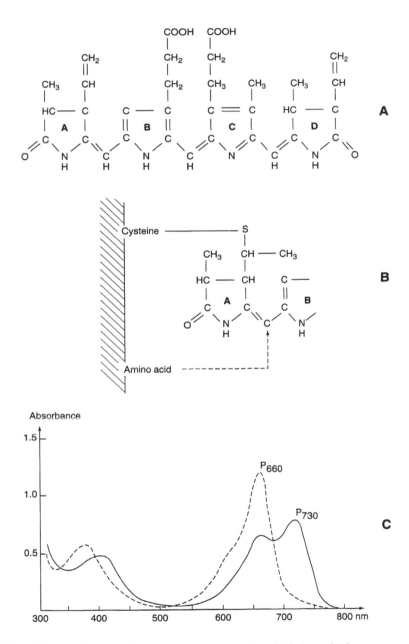

Fig. 1.15. (A) Diagram of a phytochrome molecule, which has a high molecular weight (245,000 dalton), the photosensitive or chromophore part of which is constituted of an open tetrapyrrolic nucleus. (B) This chromophore is linked to proteins of 1100 amino acids by the intermediary of a cysteine in position 321. (C) This molecule presents two absorption spectra of reduced form (P_{730}) and oxidized form (P_{660}).

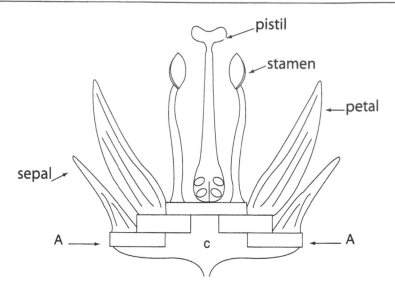

Fig. 1.16. Diagram of location of three types of genes controlling floral ontogenesis. These genes are arranged concentrically (bottom of figure) and control the realization of four whorls of floral organs (top).

involved in replacing the expected organ by another organ normally differentiated on another whorl (e.g., differentiation of stamens when sepals are expected). The discovery that some of the genes that control the differentiation of floral organs are "homeotic", i.e., have a "homeobox", which codes for transcription factors controlling the activity of other genes, has only confirmed this parallel. The mutations may be simple, but others seem more complex (double or triple mutants). An "*apetala*" mutation has been described, for example, characterized by the presence of only petals and carpels in the flower. A mutation called "*pistillata*" has flowers formed of sepals and carpels but in double the quantity found in normal flowers. A third mutation called "*agamous*" corresponds to flowers that have only sepals and petals without any real reproductive organ. Some double mutants differentiate only carpels, others only sepals. Triple mutants replace all their floral parts by organs comparable to leaves. These observations are summarized in Fig. 1.17.

Coen and Carpenter (1993) and Weigel and Meyerowitz (1994) proposed an original interpretation of genomic controls of the floral edifice. The four floral whorls that differentiate the organs (sepal, petal, stamen, and carpel) are placed under the influence of three groups of genes having exclusive functions in some cases and combined ones in other cases.

According to what is now known as the theory of Coen and Meyerowitz, the four whorls of floral organs are dependent on the

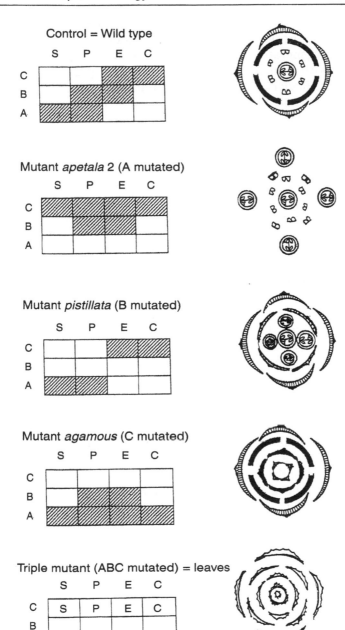

Fig. 1.17. Table summarizing the floral diagrams of *Arabidopsis*, of three simple mutants and a triple mutant. At left are the respective activities corresponding to normal or mutated genes, as interpreted by Coen and Meyerowitz.

regions A, B, and C, the domains of action of three classes of homeotic genes controlling the development of floral organs. Region A can express alone and thus generate sepals or combine with B to yield petals. C can be combined with B to organize the stamens or express alone and thus initiate carpels. Regions A and C are mutually exclusive and the absence of expression of one following a mutation leads necessarily to the expression of the other.

The ABC genes code for the synthesis of transcription factors comprising a well-conserved proteic domain, the MADS-box, which allows these proteins to interact with their target DNA sequences.

The ABC model of genomic control of flower development has been completed by the discovery of two supplementary pathways: pathway D, which controls the development of ovules within carpels, and pathway E, described for the first time in 2001, which must express at the same time as B, C, and D for the development of the three innermost whorls of the flower. In this way, the formation of stamens can be provoked by overexpressing the genes B+C+E by transgenesis.

Experiments were designed to verify the stated propositions by means of specific radioactive mRNA probe of each of the genes implicated in the development of floral organs. It has been demonstrated, for example, that the probe recognizing the RNA formed by the *pistillata* gene is fixed in the lateral regions of the floral meristem, precisely where the petals and stamens differentiate, while the RNA corresponding to the *agamous* gene are found in a much more axial region where the stamens and carpels develop (Fig. 1.18).

It should be noted that the diagram proposed seems to adapt itself well to two types of flower: *Arabidopsis* (Brassicaceae) and *Antirrhinum* (Scrofulariaceae), the only genera in which the number of floral mutants recorded is sufficiently high. However, these are two peculiar types of flowers and there have not yet been enough studies to conclude that all the floral types can fit within this diagram.

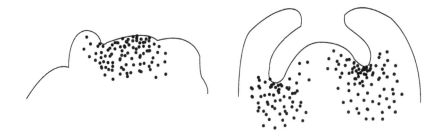

Fig. 1.18. Respective locations of specific probes of mRNA corresponding to the expression of genes *apetala* (right) and *agamous* (left) at the floral meristem.

All the events that lead from one sporophytic generation to another between which the gametophytic phase is inserted occur in the flower. The successive steps are the following:

- meiosis;

- development of male and female gametophytes;

- differentiation of the two types of gametes;

- fertilization, which is "double" in the Angiosperms;

- beginning of embryo development.

In the male sex, *meiosis* occurs in the anthers, within anther lobes, in a tissue called the *sporogenic tissue*. The cellular events in the sporogenic tissue are of the classic type and meiotic division is symmetrical, i.e., it ends in the formation of four equivalent cells having half the initial number of chromosomes and half the quantity of DNA. These are microspores (Fig. 1.19). They depend greatly for their nutrient supply on "tapetum cells", a tissue that surrounds them and in which any anomaly of function has an effect on their differentiation. This explains the existence of certain forms of male sterility that may present a handicap for reproductive performance of the plant but are actively sought after by agronomists and biotechnologists, as we will see later. As with any meiotic event, there is a segregation of parental chromosomes and consequently the distribution of alleles varies from one microspore to another. This is one of the elements of genetic mixing that sexual reproduction leads to.

In the female sex, meiosis is reserved solely for one of the cells of the sporogenous tissue called the *nucellus*. This tissue represents the central part of the ovule, protected by the integuments but in contact with the inner environment of the ovary by the intermediary of a "small door", the *micropyle*. Here the meiosis is asymmetrical and most frequently only one of the descendants, the *megaspore*, conserves the entire cytoplasmic heritage. The other three, which are disadvantaged, quickly die, constituting a sort of provisional crown for the survivor before disappearing completely (Fig. 1.20).

Let us note that, unlike the microspores, which are accessible for a possible culture and even easily accessible in some species, the megaspore is deeply embedded and closely linked with the surrounding nucellus and it is difficult or even impossible to reach.

The development of male and female gametophytes occurs by the intermediary of "haploid mitoses" that constitute a specificity of the plant world. Much more numerous in the ferns to constitute an autonomous chlorophyllian organism, they are found in a smaller number in the higher plants, and gametophytes remain discrete parasites of the sporophyte. This is a delicate period for the plant

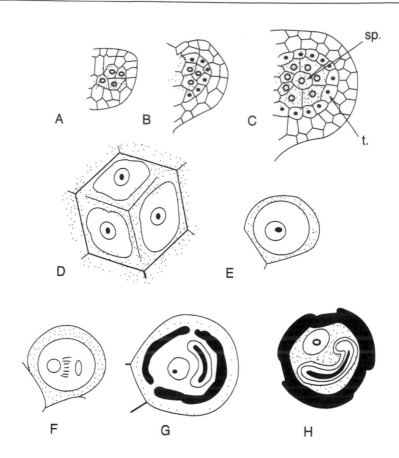

Fig. 1.19. Origin of microspores and transformation of them into male gametophyte or pollen grains. Within the anther, the sporogenous tissue (sp) develops following several mitoses (A and B). Receiving their nutrients from tapetum cells (t) (C), the sporogenous cells undergo meiosis to result in groups of four microspores (D). These become isolated (E) and are transformed, undergoing a mitosis and thickening their wall, into pollen grains (F to H).

because the microspores and megaspores have two complete ontogenic programmes in their genome and can use only one, the *gametophytic ontogenic programme*. We will see that it is now possible, thanks to biotechnology, to implement the other programme, normally silenced, which leads to the development of androgenetic or gynogenetic haploid sporophytes. These sporophytes are of great interest in agronomy.

The male gametophyte of Angiosperms corresponds to the *pollen grain*; the female gametophyte is the *embryo sac*.

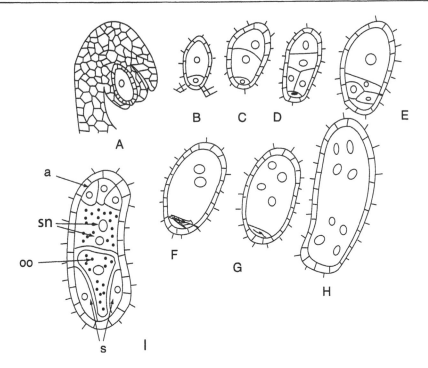

Fig. 1.20. Origin of the macrospore and transformation of it into a female gametophyte or embryo sac. Within a young ovule (A), a cell of the nucellus divides unequally (B). The larger cell, which is the archaespore, undergoes an asymmetrical meiotic division (C, D, E). Three of these cells necrose while the fourth, the largest and inheritor of most of the cytoplasmic mass, becomes the megaspore at the origin of the female gametophyte: the embryo sac. The formation of this gametophyte comprises three successive karyokineses at the origin of a syncitial structure with eight nuclei (F, G, H). Divisions appear and realize the classical structure of the sac with its two synergids (s), three antipodes (a), oosphere (oo), and two secondary or polar nuclei (sn).

The male gametophytic programme occurs synchronously in all the microspores. A preliminary karyokinesis not followed by cytokinesis separates the *reproductive nucleus* from the *vegetative nucleus* or *tube nucleus*, so called because it directs the growth of the pollen tube during the germination of the pollen grain. During this time, large secretions of the parietal material thicken the protective layers of the pollen grain. The reproductive nucleus is surrounded by a small cytoplasmic territory that has some plastids and mitochondria and is separated from the territory of the plant cell by a simple profile of endoplasmic reticulum. It initiates a new cycle of division but, depending on the species, this cycle runs until the telophase that separates two sperm nuclei (tricellular pollen) or may even stop in the prophase, which is late in reaching the tip of the pollen tube, just before it enters the ovule (bicellular pollen). These two sperm nuclei will have two different destinies: one will fertilize the oosphere to produce the zygote, while the other will be the origin of the endosperm with the two secondary nuclei of the embryo sac. Recent studies seem to indicate that their destinies are programmed, perhaps by the cytoplasmic environment: one controls most of the organelles belonging to the category of mitochondria and the other controls the plastids. These differences of environment will also allow them to be separated by a differential configuration.

The female ontogenic programme leads, through three successive karyokineses, to the development of an embryo sac, a syncitial with eight nuclei, lodged in a cavity of the nucellus limited by the remains of the megasporal wall. Some walls will appear and replace the syncitium by a cellularized structure, except for two of the nuclei, which will persist to share an individual territory in the central part—the secondary nuclei that will participate in the development of the endosperm. The other cells will see a differentiation and a role assigned by their position within the sac: antipodes (generally three) at one of the poles, two synergids and the oosphere at the other pole. The oosphere is a polarized, differentiated cell with a filiform apparatus and the presence of a large vacuole, which is rather rare for a gametic cell.

In plants, fertilization seems to be the conclusion of a great adventure in that there are numerous obstacles to be overcome and conditions to be met. The encounter of gametes is a delicate combination of risk, prediction, and programming, a series of narrow and highly specific targets that is, however, accompanied by strategies to overcome or avoid obstacles.

The first step of this adventure is *pollination*, which determines the landing of a pollen grain on the surface of the stigma. Pollination can occur by means of wind (anemophily), water (hydrophily), insects (entomophily), birds (ornithophily), or other animal vectors. It could be effected within the same flower (self-pollen) or between flowers, whether they belong to the same plant or different plants (allopollen). Cross-pollination is obviously the rule for *dioecious plants* (those with separate sexes) or even for *dichogamous plants* (the two sexes do not reach at maturity at the same time). Self-pollination is essential for *cleistogamous plants* (flower closed at maturity).

The second step for the pollen grain to cross is the cuticle of the stigma. Plants have developed a highly complex system of recognition that is similar to the immune system of higher animals, with mechanisms of acceptance and rejection (Fig. 1.21). The result is the germination or non-germination of the pollen grain. It could also be that there is a beginning of germination that allows the pollen tube to cross the cells of the stigma but its growth is found to be blocked a little later at

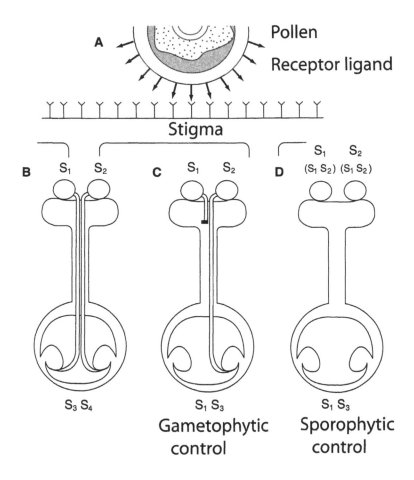

Fig. 1.21. Pollen self-incompatibility. Plants have various ways of controlling pollination. Pollen incompatibility seems to be the most elaborate because it relies on a virtual immunological reaction between the pollen grain and the cells of the stigma that triggers a multigene complex called S (for self-incompatibility). The presence of an allele common to pollen and stigma triggers the rejection reaction. Glycoproteins play the role of intermediary. The growth of the tube is found to be interrupted, either from the stigma (D) or later during the progression of the tube in the style towards the ovule (C). This blockage causes the intervention of the formation of dimeric proteins and the secretion of a particular sugar, callose. The absence of a common allele authorizes the movements of water and activation of enzymes such as cutinases (A, B). The expression of the S genome is controlled simply by the haploid genome (gametophytic control), as in the Solanaceae, or by the diploid genome (sporophytic control), as in the Brassicaceae. In this latter case the pollen grain receives proteins from the diploid tapetum cells during its differentiation. S_1, S_2, S_3, S_4, alleles of the S gene.

the style cells. The physiological control seems to act through various mechanisms, particularly movements of water from the stigma towards the pollen grain, the appearance or otherwise of a secretion of callose, a very specific sugar at the surface of the stigma, and the activation of certain enzymes including a cutinase that allows the pollen to hydrolyse the cuticle of epidermal cells. The recognition itself occurs by means of proteins belonging to the two partners, pollen and stigma, which could form a dimeric structure when their three-dimensional organizations are complementary. These three events, movements of water, secretion of callose, and enzymatic activations, are also closely linked. The physiological mechanisms are themselves controlled by genes belonging to a group particularly rich in alleles, the S genes, so called because they control pollen *self-compatibility*. These genes could be expressed in the haploid state (gametophytic control) or diploid state (sporophytic control). Various hypotheses have been proposed, as summarized in Fig. 1.21. Control by compatibility/incompatibility has a considerable importance with respect to the reproductive strategy of the plant—autogamy or allogamy—and on the genetic potential of later generations.

When all the controls are exerted, the pollen grain is authorized to grow within the cells of the style up to the funicle, which connects the ovule with the wall of the carpel. By this means the pollen grain reaches the ovule. The end of the tube opens and liberates the two spermatic nuclei in contact with the upper part of the embryo sac, often even at the level of a synergid that is destroyed by this irruption of the tube.

One of the nuclei, the one that tends to be surrounded mostly by plastids, fuses with the oosphere to form the zygote, while the second, in which the surrounding cytoplasm is rich in mitochondria, fuses with the two central nuclei to give rise to a cell that is the source of endosperm.

From the double fertilization that occurs, two different ontogenic programmes of development are initiated simultaneously: that of the endosperm and that of the zygotic embryo. Development of the endosperm occurs often at the outset in a syncitial form, and cellularization appears only later and sometimes incompletely. The role of the endosperm is, in most cases, purely nutritive for the zygotic embryo but it also serves sometimes as a transfer tissue for nutrients between the nucellus, which it directs, and the embryo. In some species, it could be present in a liquid form (coconut milk, for example) or even in a *Haemanthus*, where it serves as a support material with numerous films on the chromosomal movements during mitosis. In the case of crosses between genetically very distant varieties, the endosperm may develop vigorously and sometimes even at the expense of the embryo, endangering it. To save the embryo, it is sometimes necessary to remove it from the influence of the endosperm by extracting it from the interior of the ovary in strictly aseptic conditions and culturing it in a suitable medium. This is a sort of "micro-caesarian". Such a *saving of embryos*, very valuable for research laboratories for varietal improvement, is one of the tasks of the plant biotechnologist.

Embryo development can be divided into several periods. *Early development* is an important period for the growth of the plant. It is marked by an exploitation of potentialities and information recorded in the female gamete during its maturation. The mitoses that occur are in fact divisions of segmentation, a re-forming of the egg cell, without actual increase in the volume of the embryo, at least at first. These divisions seem to follow particular mechanisms because they seem less sensitive to antimitotic drugs. Some observations seem to indicate that they are *orderly*, i.e., the migration of chromosomes is not entirely random but takes into account their parental origin and the nature of the DNA molecule that constitutes them, whether original or copy. They are often asymmetrical, engaging sister cells in very different destinies. This is how a cell that is the origin of the *suspensor* is established. This first organ, located near the micropyle, is made up of a file of cells that play an important role in the trophic relationships between the mother plant and the developing embryo. The sister cell divides to result in a globular embryo, the morula stage, sometimes called the *proembryo*. The planes of symmetry of the embryo appear only later. It is perfectly bilateral in the Dicotyledons following the simultaneous and symmetrical development of two cotyledons on either side of the young *shoot meristem* (Fig. 1.2). The axis of symmetry is materialized at its two poles by the development of *shoot* and *root* meristems. The symmetry is less clear in the Monocotyledons following the development of a single cotyledon. Several hypotheses have been proposed to interpret the peculiarities of development of Monocotyledons, some based on evolutionary considerations and others based on the observation of mutants of development.

At the same time that the embryo begins to develop, the ovule is charged with reserves and is transformed into a *seed*, i.e., an organism of resistance that could, independently of the mother plant, survive a short or long period (sometimes of several years!) and a more or less unfavourable season (severe cold, drought). This transformation occurs according to various modalities that lead to varieties of seeds. However, in all of them there is a thickening and lignification of the integuments, accumulation of sugars (mostly starch), lipids, proteins, and tannins, as well as pronounced dehydration that blocks most enzymatic activities. The reserves may be stored in the endosperm (seed with endosperm), in a surviving part of the nucellus (seed with perisperm), or in the embryo itself, particularly in the cotyledons (seed without endosperm). The reserve products may separate spontaneously depending on their chemical family (*globoid* and *crystalloid*), the whole constituting the *aleurone grain*. This development phase corresponds to *late embryogenesis*.

From the economic point of view, the grain or seed plays a considerable role in agronomy, which we will take up in greater detail.

As the ovule becomes a seed, the *ovary is transformed into a fruit*. Here also, there are many modalities leading to a large variety of fruits—dry dehiscent or indehiscent fruits (capsules, follicles, pods, achenes) or fleshy fruits (berries, drupes, composite fruits)—several examples of which are given in Fig. 1.22. These are structures designed for dissemination of plants via transport strategies called *chories*: *anemochory* is transport by wind, *hydrochory* is transport by water, and *zoochory* is transport by animals. If the seed is transported only by its own weight, its strategy is called *barochory*, which first led Newton to discover universal gravity. These variants are all considered to have the same objective: to help the species conquer space and time.

In most species, the seed becomes dormant. This could be for a short or long period and the lifting of dormancy requires a certain period of exposure to cold, or *vernalization*. The saturation of the seed also plays an important role in the resumption of embryo development, which is noted at the growth of the rootlet and the elongation of the stemlet. The embryo is thus transformed into a young plantlet.

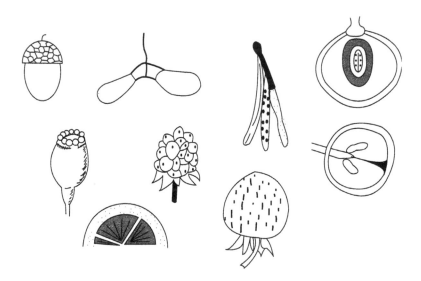

Fig. 1.22. Some examples of fruits. From left to right, the acorn, the samara of the maple, the siliqua of *Arabidopsis*, section of a drupe, the pixidium of poppy, the composite fruit of mulberry, section of a grape, section of a citrus, and the false fruit of the strawberry plant.

c) Asexual reproduction

Sexual reproduction can be short-circuited following an accidental or in some cases programmed event. Various processes could result in the development of a new living organism without genomic combination. The classic case is parthenogenetic development of the oosphere, i.e., without the intervention of the male gamete. There are also cases characterized by the development of one of the haploid cells of the embryo sac, a phenomenon referred to as *apogamy* or apogamous development. The embryos that form under these conditions are in principle haploid. In reality, regulation phenomena (karyokinesis followed by nuclear fusion) often re-establish diploidy and thus the potential fertility of the plant. *Apomeiosis* is defined as the development of a cell of the nucellus that does not undergo meiosis and thus directly results in a diploid and consequently fertile embryo. Apomeiosis and apogamy are sometimes combined under the general term *apomixis*. All these spontaneous and abnormal developments are involved in the process of so-called *natural parthenogenesis*. Biotechnologists are interested in the potential of *artificial parthenogenesis*, which plants have to widely varying extents. Microspores, for example, cultured under certain conditions, could develop into a haploid organism. This is called *androgenesis*. Similarly, an ovule cultured *in vitro* could lead to the development of an oosphere by *gynogenesis*. The *bulbosum effect* can also be considered a kind of parthenogenesis: biotechnologists sometimes carry out a simple activation of oosphere development in a plant such as wheat, after pseudo-fertilization by a male gamete of barley (*Hordeum bulbosum*) in which the chromosomes are subsequently eliminated during early development. The wheat plants that result are thus haploid.

d) Vegetative propagation

Vegetative propagation represents a third form of plant propagation and colonization of space. Already highly developed by nature in the Thallophytes, in the form of propagules, soredia, and hormogonia, they are as active as sexual reproduction in the extension of populations of Cormophytes. Vegetative propagation may be *natural* or *artificial*. It is, in the latter case, particularly valued and exploited in plant biotechnology, as we will see later. Most people are familiar with the stolon of the strawberry plant, which is a stem growing in a plagiotropic fashion, i.e., parallel to the soil, and which continually extends to differentiate buds that will give rise to new individuals. There are also the suckers of poplars and other trees, which differentiate from roots. Bulbs and bulbils represent a form of propagation found in many Liliaceae. *Bryophyllum* develops epiphyllus buds that fall on the soil and

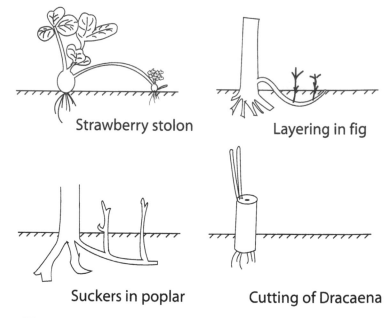

Fig. 1.23. Some classical examples of natural vegetative propagation.

rapidly give rise to new plants. The potato tuber is an example of tuberized rhizome involved in the propagation of individuals (Fig. 1.23).

Propagation by cuttings is rare in nature but is widely used artificially, in arboriculture and horticulture. This technique uses the fundamental potentialities of plant cells to dedifferentiate and redifferentiate, notably to establish a root system. Artificial layering reproduces, in many plants, the natural runners of brambles or fig trees. Grafting can also be considered a particular form of artificial vegetative propagation. *In vitro* culture of cells or tissues and the regeneration of plants from these explants, which leads to one of its chief applications, *micropropagation*, is an integral part of processes of vegetative propagation, and on a large scale. We will look at it in detail in the second part of this book.

1.2. AGRONOMY

1.2.1. Wild plants and cultivated plants

Plants that are found spontaneously in nature, in conditions in which they appear to have always lived and to which they are consequently well adapted, are called *wild plants*. Plants that are introduced and maintained by humans in artificial environmental conditions for profit are called *cultivated plants*. The distinction is not always easy to make because plants introduced at certain periods of history have been able to adapt perfectly to their new conditions and form large populations. One example is the small duckweed from Canada, which has invaded water bodies throughout the world. On the other hand, some wild plants that were once widespread presently constitute only relic populations. For example, the beech tree used to cover immense territories and today survives only in very limited areas (for example, in the Sainte-Beaume Massif in the South of France and in other dry areas). Some wild plants require and are adapted to microclimates and form only small colonies that do not attempt to expand. These are *endemic plants*.

In addition to their adaptation to climatic conditions, wild plants are generally less sensitive to diseases and parasites. Their hardiness makes them reservoirs of *resistance genes* for the plant pathologist and the biotechnologist. Since they sometimes compete with cultivated plants for occupation of a given territory, they are also considered "weeds" by the farmer and agronomist. These notions of good and bad plants are highly subjective: the renowned botanist Yves Baron has said that, in a field, wheat could be considered a weed that hampers the growth of poppy populations. The "treatment" of weeds using *herbicides* is one task of agriculture. However, eradication of some of these "weeds" deprives the agronomist of a useful reserve of resistance genes, which is why they need to be protected by special measures or laws or conserved in *botanical conservatories* to constitute gene banks. What would have happened if the weed *Arabidopsis*, which today reigns in all the laboratories of plant physiology and genetics, had been exterminated?

1.2.2. Agriculture and agronomy

In the control of plant populations, the term *agriculture* has an *empirical* connotation while *agronomy* has a *scientific* reputation. The recent evolution of cultivation practices tends today to link these two approaches closely, even though agriculture seems to be the field application of principles and ideas of agronomy.

The early civilizations practised hunting, fishing, and gathering, and the protection of resources was ensured by a form of *nomadism*. The first civilizations of farmers appeared in the Middle East in the Tigris and the Euphrates river valleys, in what is today Iraq, towards the 9th century BCE. From this centre (or three, according to some historians), the cultivation of certain edible plants spread and developed in three directions: Asia, across Egypt to the Magreb, and Europe (Fig. 1.24). It is likely that there were also centres in Central and South America. This cultivation was accompanied by a *sedentarization of populations* and territorial advances occurred through movement of descendants. Some

Fig. 1.24. Simplified map of the progression of agriculture into Europe, the Magrab, and the East from centres in the Middle East.

estimate these "movements" at around 20 km per century; others suggest they were accelerated by periodic and successive invasions of a much less pacific nature. Agriculture and its train of clearing and civilization thus gradually reached western Europe. Some plants survived migrations but they were also enriched by the input of new local species. The domestication of the horse towards the fifth millennium BCE brought about a considerable progress that sometimes has been considered as important as the invention of the harness in the Middle Ages and that of the tractor in the early 20th century. Farmers felt the need to choose the plants that seemed to yield most or if possible to modify them over generations to improve their performance. Thus emerged *mass selection* carried out within plant populations and leading to better yield (Fig. 1.25). Farmers also sought to improve cultivation practices in order to prevent soil degradation, by crop rotation and the input of manure. All these processes evolved by trial and error; they were thus *empirical* and often transmitted by oral tradition. There are of course a certain number of treatises that have marked out the agricultural history of the world, more or less well documented, commented, and reasoned.

Agriculture acquired a scientific dimension only with the emergence of two disciplines. The first to appear was *plant physiology*, particularly the study of plant nutrition and reproduction, which led to a considerable improvement in cultivation practices, harvests, and the yield of cultivated plants. The second was *plant genetics*, initiated in the

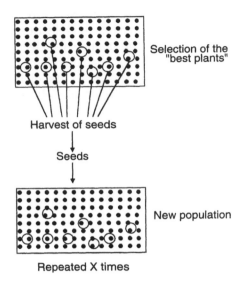

Selection of the
"best plants"

Harvest of seeds

Seeds

New population

Repeated X times

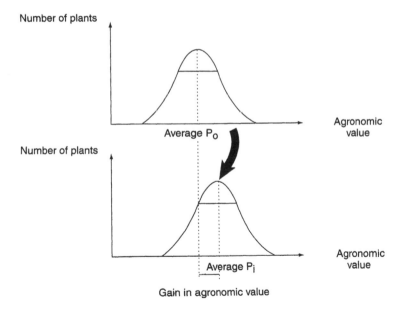

Number of plants

Average P_o

Agronomic
value

Number of plants

Average P_i

Agronomic
value

Gain in agronomic value

Fig. 1.25. The modalities of mass selection from populations by simple representative choices of the "best" individuals within a population. The criteria used are often partly subjective (size, habit, proportionality of plants). This empirical technique has resulted in considerable advances. P_o = original population. P_i = new population after x mass selections.

mid-19th century by the Czech monk Gregor Mendel, and complemented by modalities of large-scale selection, no longer over a population but of an individual (genealogical selection, backcrossing). This scientific agriculture combining the knowledge of physiology and genetics became *agronomy*, a science with its own principles, rules, practitioners, engineers, breeders, and technicians. Agronomy today maintains its scientific character by using the most recent techniques of genetic engineering and biotechnology. The application of these new technologies to agronomy is the main subject of this book.

1.2.3. The objectives of agriculture

The essential missions of agriculture are the following:
- to ensure optimal environmental conditions for the cultivated plant;
- to best exploit the intrinsic potential of the plant;
- to protect and if necessary care for the plant;
- to market the plant and the products made from it.

a) The environment of the plant

Farmers are concerned with the physical, chemical, and biological environment of the plants they cultivate. The *physical parameters of the plant environment* concern the practice of ploughing (working and aeration of soil) to ensure a suitable soil structure and texture, support for the plant, as well as a capacity to provide the correct mineral and water supply for it. Management of the quantity of water available for the plant (by irrigation or drainage) is often a major concern for the farmer. The quality and quantity of light must be ensured by the choice of a favourable orientation and the absence of obstacles so that the plant can benefit from an insolation adapted to its needs.

The *chemical environment* of the plant depends on elements naturally available in the soil or resulting from photosynthesis, as well as inputs of micro-, macro-, and trace elements from fertilizers and additives. The relative levels of nitrogen, phosphorus, and potassium (NPK) in these fertilizers is fundamental. With respect to the *biological environment* of plants, the essential factor is competition from other plants.

For wild plants, the *density of plant cover* is naturally controlled by the secretion of substances toxic to other plants. For example, the juglones secreted by the walnut tree prevent other plants, including its own seeds, from growing. Such an effect at distance is called a *teletoxic effect*. For cultivated plants, the farmer controls the plant cover by planting at a particular density (most often seed density) and eliminating unwanted plants by mechanical methods (hoeing) or by the use of herbicides. Herbicides are more or less specific and are usually solutions of synthetic hormones at particular dosages. 2,4-D and some of its derivatives are most commonly used to eliminate dicotyledons from fields planted with cereal crops.

b) Exploitation of the potentialities of the plant

All plants do not have the same agronomic potentiality. They are generally chosen on the basis of the major products that can be extracted from them: e.g., plants that yield sugar and flour (principally starch), oilseed crops (oleaginous), plants rich in proteins (proteaginous). Some plants, such as soybean, are cultivated for oil and protein (oleoproteaginous) because they have a relatively balanced composition of the two products. These relative levels in cultivated plants can be diagrammed and positioned in a pyramid as shown in Fig. 1.26.

Plants with principally sugar and starch content were the first to be selected for yield and quality (beet, cane). They were also the first to reach a level of self-sufficiency in this sector. Since then, chronic overproduction has resulted in the search for industrial uses, particularly in the field of energy (carburol from alcohol). Breeding programmes subsequently addressed oilseed crops and Europe became self-sufficient in production a few decades ago. Since then, the production of some surpluses has also motivated producers to aim for industrial uses, particularly towards renewable energy sources (diester). At present, the cultivation of proteaginous crops, although it has been the focus of development programmes, has not yet led to protein self-sufficiency for countries in western Europe. Those countries remain highly dependent on imports from the Americas for animal feed. These imports also contribute a great deal to the trade deficit. The same is true of a non-food plant product, cellulose.

The *reproductive capacities* of plants are a second criterion of plant selection. They essentially involve reproduction by sexual means and, consequently, the capacity to produce grain. Some grain is preserved as seed, and most is meant for food and industrial uses. The quantity of grain harvested over a hectare often represents the *yield* and subsequently the profitability of the cultivation. With respect to the natural strategy of fertilization that they have adopted, plants are classified into two major agronomic categories: autogams and allogams. Autogams (to which most grasses belong) tend to constitute homogeneous populations; autogamy leads naturally toward homozygosity. Homogeneity of the population is favourable to mechanization of farming (synchronous maturity, standard size) and is thus sought after by the farmer. Such populations are, however, threatened by the usual consequences of *inbreeding*, characterized by

Agro-food sector

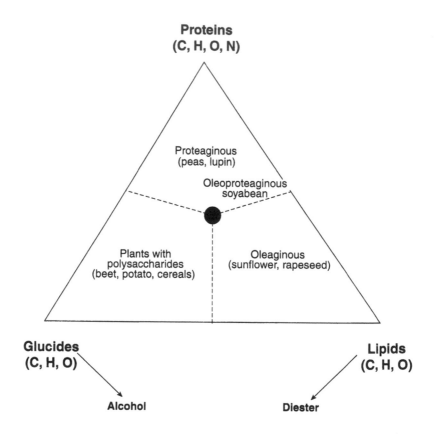

Energy sector

Fig. 1.26. Pyramid of intensively cultivated plants as a function of their glucide, lipid, and protein content. Soybean occupies a high position because of its balanced composition of lipids and proteins.

loss of vigour after a few generations. The appearance of an increasing number of mutations that stabilize in the homozygous state is the explanation most often given. The second category, containing allogamous plants such as lucerne, are the opposite in agronomic terms. The successive crosses maintain a high heterozygosity in populations, which therefore show a certain degree of heterogeneity and thus variability. This constant hybridization favours the appearance of *heterosis* phenomena, i.e., the acquisition by the descendants of qualities superior to those possessed by the parents. The greater the distance from the parents, the greater the heterosis manifested in the descendants. The heterogeneity of these populations could be a disadvantage for the farmer because it makes mechanized cultivation and harvest more difficult. On the other hand, it could be a quality for horticulturists and fruit growers, who wish to market their harvest over a longer period of time. It should be noted that within a single species there may be allogamous genotypes or varieties and autogamous ones. In peas, for example, there are varieties that have adopted the two opposite strategies: the varieties cultivated for extraction of flour produced for animal feed are autogamous and allow large-scale cultivation and mechanized harvest, while the horticultural varieties are allogamous and the gardener values being able to harvest daily a quantity just sufficient for household consumption, over many days.

To put it simply, we can conclude that there are homogeneous, stable plants that are useful for agriculture but the future of which is compromised as well as heterogeneous, unstable plants with greater evolutionary potential.

In fact, the properties of autogamy and allogamy are not always entirely exclusive. An autogamous plant may allow some pollen tubes to overcome the barriers of specificity expressed at the stigma and thus end in an allogamous fertilization. The same is true, but in a symmetrical fashion, for allogams. It is this small exception to the rule that allows the agronomist to intervene in the modalities of reproduction and carry out plant breeding programmes. These techniques of improvement are no longer *empirical*, like mass selection over populations, but *scientific*, because they are based on predictions derived from the laws of Mendelian genetics.

The selection protocol generally begins with a cross between two plants that constitute the parental generation (Fig. 1.27). One is considered *paternal* because it provides the pollen, and the other, on which the fruit is harvested, is the *maternal* plant. It should be noted that in fact in most cases symmetrical crossing is possible (except in plants with separate sexes or dioecious plants and in male-sterile plants). These crosses give rise to an F_1 generation that is homogeneous no matter what the genotypic distance of the parents. These can be crossed with each other to produce an F_2 generation or crossed with one of the parents to obtain a backcross generation. The F_2 results from a disjunction of characters and thus constitutes a population that could be heterogeneous again, but repeated crosses according to the same pattern and giving successive generations F_3, F_4, F_5, etc. will help stabilize the characters while moving the generations towards homozygosity. Agronomists generally estimate that 8 to 10 generations constitute a

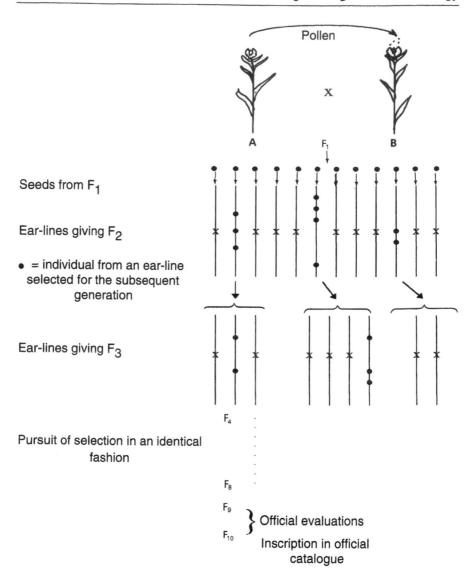

Seeds from F_1

Ear-lines giving F_2

● = individual from an ear-line
selected for the subsequent
generation

Ear-lines giving F_3

Pursuit of selection in an identical
fashion

Official evaluations

Inscription in official
catalogue

Fig. 1.27. Pedigree selection. The operation always begins with a cross
between two well-documented individuals valued for their agronomic
qualities, which results in a homogeneous F_1. These individuals are
then crossed together to result in an F_2, which restores some
heterogeneity following the disjunction of characters during the
formation of gametes in the F_1. The selection from agronomic criteria
begins with F_2. This selection is generally carried out in the same way
until the eighth generation. Then two generations of certification are
required (F_9 and F_{10}) for the variety thus created to be recognized
(inscribed in an official catalogue).

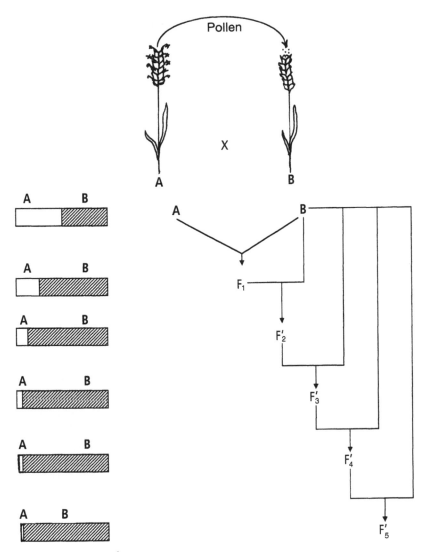

Relative percentage of
genomes A and B
in each generation

Fig. 1.28. A backcross protocol. Here the crossing of two lines A and B gives an F_1 in which the individuals, during anthesis, are crossed with the parent B and the same is done with each generation. At left are shown the relative proportions of the two genomes in the individuals of each generation. At the fifth generation, the theoretical percentage of the genome of A is no more than around 2%. The genes of A are thus said to be introgressed into B.

guarantee of stability. The backcrosses (Fig. 1.28) are also highly interesting because they allow us to find, if the selection is made well downstream, most of the qualities of the parent chosen to which are added a small number, or even one, of the qualities of the other. This is essentially a transfer of the character of one of the parents into the genome of the other, an *introgression* of the character thus selected. It is a long and delicate selection technique and we will see how biotechnologists have proposed a much quicker and more direct solution.

One difficulty that often appears with prolific species (Poaceae, Solanaceae, Brassicaceae) comes from the number of descendants to be followed, which increases exponentially with each generation. It is sometimes advisable to reduce this number and conserve only a small number of grains, sometimes only one, belonging to each generation from the F_3 onward. Breeders are certainly aware that they thus introduce chance or subjectivity into their selection, but this decision allows them to manage populations of more acceptable size. The method is known as single seed descendant or SSD selection (Fig. 1.29).

There are other selection techniques, more or less complex, sometimes better adapted to autogams than to allogams or the other way around (bulk method, pedigree selection, recurrent, reciprocal recurrent, with full-sib or half-sib tests). The study of these techniques goes beyond the scope of our discussion and the reader can refer to the works listed at the end of the chapter.

For the variety thus created to be validated and for the creator to be recognized as breeder (it is listed in the official national or international catalogue of varieties), it must undergo two successive years of assays, in four different environmental conditions (places, climates), during which it must be *distinguished* from other listed varieties, prove its *homogeneity* and *stability*, and express the agronomic character for which its inscription was requested, at a value higher than that of the control (103%).

The capacity of plants for vegetative propagation is also of interest for agriculture. This property, which involves no genetic recombination of the kind resulting from sexual reproduction, allows the constitution of particularly homogeneous populations apt to invade spaces available in the immediate vicinity. It uses the "totipotency" of somatic plant cells that, when put in new environmental conditions, are always ready to engage in a process of dedifferentiation and redifferentiation. The modalities of expression of this property are multiple and varied. Along with the examples of natural vegetative propagation already cited, we should add the multiple possibilities of *in vitro* clonal propagation or *micropropagation*, which will be studied in the next chapter. These modalities of true multiplication lead to the establishment of clones. Nevertheless, studies on growth parameters (*Phyllanthus*, for example)

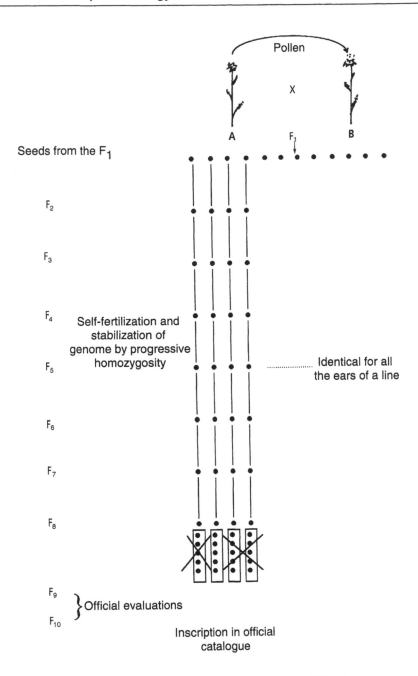

Fig. 1.29. Variant of pedigree selection leading to an SSD selection. From the F_2 of particularly prolific species, only one seed of each generation is retained. By autogamy, each generation developed from this single seed increases its rate of homozygosity.

or flowering (pineapple) of some of these clones have shown that it is possible to record a small range of variability called "somaclonal variations", the origin of which is not well understood but could correspond to individual modifications during ploidy.

c) Protection of plants, crops, and their environment

The third objective of agriculture involves the protection of cultivated plants and their environment. The dangers could come from the climate (frost, flood, drought), trauma (breakages or injuries inflicted by people and animals), chemicals (pollution), or biological sources (bacteria, fungi, insects). Several threats from various origins could also accumulate to cause the withering of the plant. For example, an injury caused by frost may allow fungi to invade the plant, carried by the sap. Or there is the damage caused by the fungus *Ceratocystis ulmi*, which is the origin of an elm disease called "graphiosis", the spread of which is facilitated by galleries dug by the larvae of the elm bark beetle.

Parasites and pests can be controlled by chemical or biological means. Chemical control is based on an arsenal of phytosanitary products including families of fungicides, herbicides, insecticides, or pesticides: organochlorines (dichlorodiphenyl trichlorethane or DDT, lindane), organophosphates (malathion, parathion), or sterol biosynthesis inhibitors (fenpropimorph, tridemorph, flusilazol, trichlopyr). Some are products designed to treat surfaces and others, called *systemic*, penetrate the tissues and are carried to the phloem for circulation, like some of the metabolites they resemble. Although the efficiency of products is not doubtful, their effect on the environment at the recommended doses is far from harmless. They may also persist and be found in undesirable levels in the water table. Some farmers have found with anguish and bitterness that their profession has become one of the major polluters of the environment.

To reduce such pollution, various attempts have been made to suppress or replace phytosanitary products with technologies closer to natural defences. The organic farming movement, for example, opposes the use of "chemical products" (but what product is not chemical?) or advocates a drastic reduction of their use. Some herbicides are photodegradable and thus do not persist in the environment. Another method is biocontrol, using parasites or predators of the pest and seeking to establish a new biological equilibrium. For example, ladybugs, natural predators of aphids, are used to control the parasitic activity of aphids on rose and other plants.

In the domain of plant protection, we will see that biotechnology has allowed several spectacular breakthroughs, some of which raise other dilemmas of ethics or civilization.

The protection of plants can also be extended to the *protection of their genetic inheritance*. Modern agriculture, in the search for high yields, has a tendency to retain only the most profitable genotypes, consequently drastically impoverishing the diversity and variability of cultivated species. Some varieties are endangered, and others have already disappeared, depriving the agronomist and the biotechnologist of an entire set of genes, sometimes valuable ones, that those varieties possessed. The need to conserve these old varieties as gene reserves or banks has led administrative and political authorities to promote the creation of *botanical conservatories* in which those varieties are protected and given particular care. Such measures have been taken in order eventually to recover the genes, which could be reintroduced into cultivated species by crosses followed by selection or directly by techniques of genetic engineering. The conservatories are specialized, some designed for fruit trees (many initiatives taken in the case of apple and pear trees), others for the Graminae, still others for some 900 varieties of potato. Plants belonging to the local flora are generally conserved in botanical gardens.

Protection of the plant's environment is also part of the mission of agriculture. Soil and water are the chief objects of attention. The disastrous consequences of deforestation and overgrazing are widely known: erosion due to indiscriminate clearing of hedgerows, and excessive salinity of some lands (for example in Spain, in the Extremadure province) in which irrigation and drainage have been poorly managed. The consequences of intensive agriculture have been no less disastrous for the water in captive phreatic layers, in which nitrates accumulate from excessive fertilizer use in intensive cultivation. These problems of water and soil protection could become major concerns for farmers of the 21st century. Here as well, biotechnology can offer a range of solutions, such as microorganisms genetically programmed for the treatment of persistent organic molecules following fertilizer and pesticide application, and thus help to decontaminate soil and water.

d) Marketing of crops

The fourth and last major mission of agriculture is the marketing of agricultural products and, consequently, the introduction of the matter thus produced into the economy of a region or country in order to perpetuate this sector. After having long been concerned with *food*, marketing of agricultural products has increasingly focused on *industry* and is today involved in practically all industrial sectors.

The *agrofood sector* remains, and probably will remain for a long time, the primary sector for marketing of agricultural products. There are, of course, frequent "crop surpluses" in the western countries that suggest that overproduction will not find an outlet. The fact that nearly half of humanity suffers from malnutrition and underfeeding and one-fourth remains at the edge of famine suggests that it may be more judicious to review the systems or routes of wealth distribution rather than to check the essential activity of agriculture. It is sometimes difficult to imagine that famines raged even in the 19th century, in Ireland for example. We must also not forget that the relative food abundance of the 1930s followed four years in which the main concern of every European was the contents of his dinner plate.

Agricultural products can be marketed at various levels of value addition. First, a *raw material* is sold directly for animal or human consumption without any transformation. This is true of fruits, vegetables that are eaten raw, bulk cereals, etc. The problems encountered are those of harvest, transport, preservation, and storage, problems that are often the focus of concern in agricultural cooperatives and the management of silos. The consumption of food may also require a minimum of transformation and conditioning—e.g., cooking, dehydration, vacuum drying—and thus enters the domain of agrofood. The next step in value addition is characterized by the need for significant transformation by a process more chemical than biological: grinding, freeze-drying, fermentation, extraction, distillation, extrusion, colouring, addition of organoleptic agents or gelling agents. Some very common products such as bread and wine undergo several operations of transformation. The same is true of food oils.

The activities of transformation of food products are already part of the industrial sector, by their technical nature and turnover. Moreover, the distinction between agrofood processing and industrial processing is often difficult to establish and artificial because the two sectors largely overlap. It is more correct to separate the food and non-food outlets of agriculture.

Non-food industrial marketing of farm products on the large scale is, overall, a vast and more recent orientation in the search for new uses for these products. There are, nevertheless, some sectors belonging to this category that are already quite well established, such as the cellulose and fibre sector in general, as well as partly the energy sector. Under the pressure of ecologists and consumers, other sectors have developed, such as packaging, plasticizers, electric and thermal insulation, largely of recyclable products, which industry is demanding in increasing quantities. Undoubtedly, this industrial sector will depend on biotechnology, in the near future and in an accelerated fashion, to find solutions in producing as well as in subsequently eliminating and recycling at lower cost.

The non-food uses of agricultural products are numerous and varied and it is not possible to enter into detail about them without compiling long, tedious lists or discussions out of proportion with the objectives of these brief summaries. For convenience, they are summarized in four categories:

- *Industrial fibres*. This category basically includes wood, cellulose and its derivatives: paper, cardboard, textiles (from cotton, flax, and hemp), plastics (remember that one of the first heat-formable plastic materials was "celluloid"). Today, these products are found in cars and in electric, heat, and sound insulators.

- *Heavy chemicals*. Products of agricultural origin are found in inks, paints, varnishes, solvents, dispersants, dyes, and detergents (soaps and washing powders).

- *Fine chemistry, plant pharmacy, and parapharmacy*. The vast majority of products with pharmacodynamic effects are known to be of plant origin (antibiotics, heterosides, alkaloids such as colchicines, digitaline, antitumorals of the vinblastine type, vincristine, taxol and other ginkgolides, perfumes, essences, and cosmetics.

- *Energy*. The combustion of wood, peat, fossil wood, and petrol constitutes the greatest part of this sector. The addition of alcohol fuels (methanol and ethanol) derived from the fermentation of sugars coming from plant matter is one of the routes chosen for the manufacture of biofuels. The oils and their derivative "diester" constitute a second route that uses the lipidic network (rapeseed oil, sunflower oil). This industry has a triple objective of reducing the consumption of non-renewable fossil fuels, contributing to the energy independence of countries with small or no fossil reserves, and reducing "crop surpluses" with a food orientation (glucides and lipids).

Despite all these present or developing breakthroughs, agriculture seems to be in a constant state of crisis. This crisis has led to the exodus towards urban centres of what at mid-century represented half the population and now represents only 6 or 7%. Agriculture operates over the long term and national or international policies are often expressed over the short term. A petrol crisis can in a few hours lead to a political or ethnic conflict, whereas a reorientation of agriculture towards the large-scale supply of biofuels will require the activity of one or several generations. There is thus a perpetual conflict in the time scales and a need for constant adaptations or reorientations. Despite this, or because of it, agriculture has a great future, associated with the development of biotechnology at its service.

1.3. CELLULAR AND
       MOLECULAR BIOLOGY

1.3.1. Information in the cell

a) Genetic information, DNA and chromatin

The concept of a cellular structure of living things is ancient but the material confirmation of this entity dates from the 16th century, with Leeuwenhoek's invention of the microscope in Holland and the observation of small cavities—cells—in cork. It was only in the last decades of the 19th century that the chromosomes were recognized in the nuclei, almost at the same time that the first laws of genetics were spelled out. In the early 20th century, the segregation of chromosomes and that of characters was correlated and thus the basis of heredity was located in the chromosomes. The discovery of nucleic acids between the two world wars as well as the identification of their organization and their properties soon after the second were the basis of the extraordinary development of cellular and then molecular biology in the second half of the 20th century. Today, it is known that it is the nucleic acid molecule, perhaps the only one designed for self-replication, that contains the information and constitutes the *genome* of the cell. The genomic information is located essentially in the nuclear compartment. A smaller but not secondary part is located in the mitochondria as well as in the plastids when the cell has them.

The DNA molecule is made up of a chain of nucleotides forming a double helix of 2 nm diameter. It is relatively long, although its length is highly variable within a cell and from one species to another. The whole DNA generally constitutes a single circle in bacteria, multiple circles in the mitochondria and plastids, and linear structures with two independent ends in the nuclei of eukaryotic cells. Each chromosome seems to contain only one DNA molecule, except when this molecule replicates itself a little before the processes of karyokinesis are initiated. In the nucleus, the DNA is associated with histone proteins to form units called *nucleosomes*. Nucleosomes form the beads in a long chain, the whole of which is called nucleofilament, with a diameter of close to 10 nm. In a nucleus, there are as many nucleofilaments as chromosomes and they are the most unrolled type of chromosomes. They can be observed only under electronic microscope and often only after metallic shading. Under a light microscope, the nuclear contents seem to be continuous and homogeneous and form the *euchromatin*, which is relatively accessible to enzymes that exploit the information to synthesize the three types of RNA (messenger, transfer, and ribosomal). In some circumstances, the nucleofilament can roll in on itself to form a sort of large cord and result in a filament of higher order called a 30 nm fibre because of its greater diameter. A combination of nucleofilaments and 30 nm fibres in a given region of the nucleus gives the chromatin of this nucleus a heterogeneous structure, which is why it is called *heterochromatin*. These filamentous structures are partly liberated in the nuclear space. Only the ends or telomeres and some regions, including the centromere, fix on the inner surface of the nuclear membrane by the intermediary of proteins that constitute a layer of varying thickness: the *lamina*. The free parts, i.e., those not linked to any support, constitute domains (Fig. 1.30) that often correspond to regions in which the information is more or less totally exploitable by transcription enzymes. These domains shorten themselves to form intensely spiralate regions around a backbone associating proteins and RNA. These structures are then found condensed at the level of chromosomal arms when these arms are organized around the prophase.

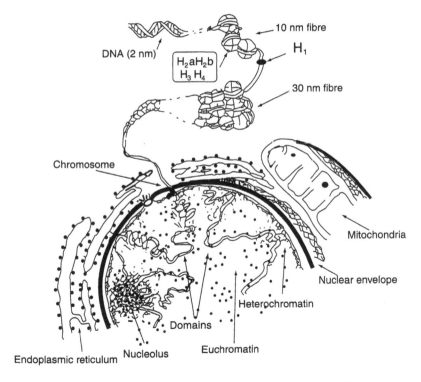

Fig. 1.30. Information within the cell. Successive cycles lead from the DNA with nucleofilament of 10 nm to the chromatid fibre of 30 nm diameter, then to the interphase chromosome with its free "domains" and regions super-coiled at the level of the heterochromatin. H_1, H_2 a and b, H_3, H_4 are histones.

The information is thus engaged in a process of replication and is no longer available for transcriptional activity.

The DNA is made up of two chains oriented in an antiparallel direction. This orientation, which determines the direction of reading for transcription as well as translation, is characterized by two asymmetrical ends called 5' and 3'.

b) The unit of genetic information: the gene

Each DNA molecule corresponds to a *linear succession of genes* controlling the activity of these genes and finally other regions, the role of which has not been determined with precision.

A gene corresponds to a succession of some hundreds or thousands of matched base pairs (bp): an average gene size in a plant such as *Arabidopsis* is 1500 to 2500 bp. Each gene comprises basically four regions:

• a promoter;
• a coding sequence;
• a terminator;
• regulator sequences.

The first three are most often continuous or, at least, contiguous on the DNA molecule; the fourth could be more or less distant and sometimes even located on another DNA molecule belonging to another chromosome (Fig. 1.31).

The *promoter* is a DNA sequence of variable size but most often having some hundreds of base pairs whose essential role is to be recognized by the RNA polymerases and to control the transcription of the coding sequence or sequences that follow. The promoters often present clusters in their sequences located in relatively constant positions. Towards 25 bp, for example, before the end of the promoter, a *TATAbox* is found, called so because it is characterized by a chain of nucleotides TATAAA. This TATAbox is itself often preceded by another box, the *CAATbox*, with a sequence CAAT, located around 75 bp. These "boxes" play a role not only in the recognition of the RNA polymerase but also in its fixation. It is at this level that particular proteins called *transcription factors* are involved. The presence and relative positions of these two boxes are particularly useful in identifying a promoter in a DNA sequence. A promoter may control several coding sequences that are contiguous or regularly spaced from one another and, consequently, the almost simultaneous synthesis of several proteins. These proteins thus often have complementary activities, as, for example, with an enzymatic protein and its permease that allows it to cross membrane barriers. This situation is principally found in bacteria. The whole is thus called an *operon* and each of its elements a *cistron*.

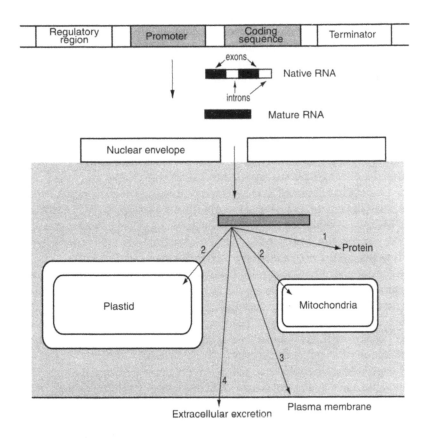

Fig. 1.31. Organization of the gene and its role in the synthesis and emission of messengers in the nucleus after maturation. These messengers migrate into the cytoplasm, where they are translated into proteins. The proteins are used *in situ* (1), exported to cytoplasmic compartments (plastids and mitochondria), incorporated in the plasma membrane (3), or exported to other cells (4).

The *coding sequence* is the essential part of the gene since it alone is found at the protein and ensures its specificity. Generally, the beginning of a coding sequence is found in the sequence of the *primary messenger RNA*, which must then undergo processes of polyadenylation and maturation. The first operation is characterized by the fixation of the 3' end of a *poly A trailer* with this messenger, by means of the intervention of a poly A-polymerase. This primary messenger RNA will undergo a phase of maturation characterized by the deletion of certain sequences called *introns*, by means of highly specific enzymes. This is the process of *excision*. The conserved fragments or *exons* are thus joined together by other enzymes; these are the processes of *splicing* or *ligation*. Normally, the mature RNA contains all the exons united by splicing but sometimes, depending on the circumstances, this mature RNA encloses only some exons and not always the same ones. Thus, a native RNA may correspond to different mature RNAs translating different proteins. This is called *alternative splicing*. The phenomenon increases the variability of expression of a DNA sequence. Such variability could be still further augmented by *editing* of RNA when certain bases are spontaneously replaced by others (adenosine replaced by inosine). The enzymatic complexes of excisions and splices, large particles, are called *snurps* or *spliceosomes*. Only the RNA that have reached maturity are thus translated in the cytoplasm by ribosomes. When copy DNA is used in place of the coding part of the gene, the maturation is not necessary because this molecule is the true reflection of exons only. In fact, this copy DNA is a synthesis of DNA realized by the intermediary of a particular enzyme extracted from certain viruses—the reverse transcriptase—and controlled by mature messenger RNA.

The coding sequence normally terminates in a *stop codon* that marks the beginning of the terminator. This codon does not correspond to the tRNA and it cannot therefore code for the coupling of any amino acid (for example UAG). There are also *non-sense codons* and *termination codons*. The terminator is, in fact, a short DNA sequence that also contains codons for polyadenylation of messenger RNA.

The *regulator sequences* are often much more difficult to determine. They may be located upstream of the same chromosome but also somewhere in the genome. They serve to adapt the level of transcription to the instantaneous needs of the cell. They confer a certain specificity of transcription to certain organs (roots, leaves, flowers) or to precise stages of development. This property can be characterized by a succession of foliar forms linked to precise stages of development as indicated in Fig. 1.32. Some genes that appear to be less or not susceptible to the processes of regulation are constitutive genes. Finally, there are isolators, sequences that can temporarily isolate and protect some regions of DNA against the influence or control of other regions of the same DNA.

c) The different types of genes

All genes are constructed on the same principle but they differ considerably in the nature and level of activity of their promoter (low, medium, high), the length of their coding sequences, and their richness in regulation sequences. We will see that it is possible to construct by genetic engineering a coding sequence, promoter, and terminator that come from three different genes. This synthetic gene could function perfectly provided the linking of elements and distances that separate them from one another are respected.

Fig. 1.32. Example of heteroblastic development in the small aquatic fern *Marsilea vestita*. During the growth of embryo and then juvenile development of the plant, there is a successive appearance of leaves that are aciculate (1), spatulate (2), bifid (3-4), quadrifid (5), expanded quadrifid (6), and sporangiferous (7) r, roots; rh, rhizome.

The total number of genes varies greatly from one species to another. There are between 4000 and 6200 genes in bacteria for a genome of 4 to 5.7 million bp, or around 6000 in baker's yeast, a genome of more than 14 million bp in which the entire sequence was mapped in 1996. The yeast thus has about 1.5 times as many genes as bacteria for a genome that is 3.5 times as long. *Arabidopsis*, which has the smallest plant genome known, has around 120 million bp. In plants, this number varies greatly if one considers the total length of the genetic information, i.e., imagining that one can link the DNA molecules of a single cell end to end, the total length of the genetic information is 440 mega bp in rice, 16,000 mega bp in wheat, and 30,000 mega bp in tulip. It should be noted that the DNA molecule of the tulip is 8 times as long as a human DNA molecule. There is thus no direct relationship between the level of evolution and the length of information.

These important differences between genome sizes often come from the degree of repetition of genes. Genes are more or less redundant. There are three categories of genes in this respect: those with a *low* degree of repetition, in which there are only a relatively small number of copies; genes with a *medium* degree of repetition, which are much more widespread; and those with a *high* degree of repetition, in which the number of copies exceeds several thousands. The last category seems to be widespread in most plant genes and it is not rare for a plant to use most effectively and in all circumstances only a small fraction of its genomic potential.

Some genes located close to one another present sequence analogies, even though these code for the synthesis of distinct products. These are gene *clusters*.

In plants, as in animals, there are genes specific to stages of development. In animals they are the source of organ differentiation: feet and antennae in insects, the limbs or eyes in vertebrates. We have earlier seen that such genes are the source of different whorls of floral parts. Others, such as *knodded* genes, are involved in the functioning of cells of the initial ring. These genes are characterized by the presence of an identical sequence of around 80 bp in their promoter, a sequence called "homeobox" because it is highly homogeneous within a single family of genes acting in a very similar fashion. These genes, which are called *homeotic genes*, have been particularly well conserved during the evolution of organisms. Recent studies seem to indicate close similarities in these families even for species that are phylogenetically very distant (e.g., the eye of the drosophila and the eye of the mammal).

One category of genes that is highly valuable in genetic engineering is that of *reporter genes*. These are genes with an activity that is relatively easy to observe in the organism in which they are expressed. Often there is a gene that codes for the synthesis of an enzymatic protein having the property of transforming a colourless substrate into a coloured reaction product. This is the case with the GUS gene, which codes for the synthesis of glucuronidase transforming certain substrates such as X Gluc (5-bromo-4-chloro-3-indolyl-D-glucuronide) into a blue precipitate of di-X-indigo that is easily detected (Fig. 1.33). In this way, the presence of the gene is easily identified at all scales of observation: macroscopically by observing the appearance of small blue patches in the tissues, as well as microscopically to allow a more precise definition of the exact location of enzymatic activity, on the scale of the cell under light microscope or on the scale of an organelle under electron microscope. When the level of activity of a promoter needs to be tested, a chimera gene is constructed by placing the GUS gene along with the promoter to be tested: the intensity of the coloration obtained is a true reflection of the degree of promoter activity. There are now several promoters available that are easy to use and are selected most often on the basis of the material used. Among the most commonly used are CaMV, NOS, genes of heat shock proteins (HSP), and SAUR gene.

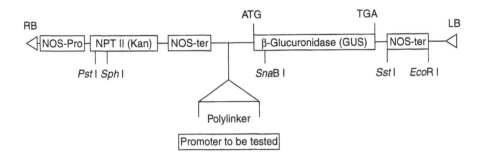

Fig. 1.33. The GUS gene is among the "reporter genes" such as lac Z genes (galactosidase, chloramphenicol acetyl transferase, neomycin transferase, or luciferase). These genes are used to label cells in which they are introduced and in which they are expressed. They can eventually be used to test the transformed cells.

1.3.2. Controlling genetic information

a) The plasticity of the DNA molecule

The control of the genetic patrimony of a plant in order to impose on it the desired form, performance, or quality is an old dream of farmers. Empirical selection and then the application of rules of a nascent genetics were two successive approaches that supported a large number of expectations and investments. Crosses, backcrosses, and selections have represented and still do represent the basic technologies essential to the control of genetic patrimony, not of an individual but of its descendants. The many-directional and spectacular advances of biology in the second half of the 20th century have placed at the agronomist's disposal an additional range of tools that are particularly effective in modifying the genetic inheritance of individuals, microorganisms at first and then plants and animals.

The mastery of the genome flows directly from the mastery of the DNA molecule. To be able to identify, extract, dissociate, fragment, reconstitute, establish the sequence of, and transfer this complex molecule from one organism to another are the elementary conditions that lead to the conception and then realization of a genome that is new because it is recomposed and often original. The tools necessary to this enterprise are, in rare cases, purely physical and, in the vast majority of cases, chemical and more precisely enzymatic.

Of the physical tools, temperature variations are the most frequently used. The two chains that make up the DNA molecule, linked to each other by hydrogen bonds, begin to separate when the temperature reaches about 60°C and are entirely separated by the time it reaches 100°C. In fact, the temperatures of separation vary locally with the respective richness in A-T bonds (two hydrogen bonds) and G-C bonds (three hydrogen bonds). Thus, a region rich in G-C will separate at a temperature that is clearly much higher than will a region rich in A-T. The temperature for which 50% of DNA becomes single-stranded is called the *fusion* temperature and is characteristic of a given type of DNA. We will see that the influence of these temperature variations is largely used in a process of *in vitro* polymerization of DNA, particularly in the technique called *polymerase chain reaction* or PCR. This technique is based on the cyclic alternation of denaturation and renaturation of DNA in which phases of polymerization are interposed, i.e., copies of DNA strands. The pH, the energy level of certain kinds of radiation (UV, for example), and the water content of the environment also influence the conformational characteristics of the DNA molecule.

The chemical products that intervene in the structure or organization of the DNA molecule are numerous and it is not possible to make an exhaustive study of them within the scope of this work. Some metals, such as nickel, could provoke base substitutions and, consequently, occasional mutations in the genomic sequences. Some antibiotics lead to the same results but by a different pathway. Others interrupt the transcription and introduce stop codons or error codons. Some substances, like ethidium bromide or EtBr, could intercalate between the two chains of DNA and compromise the fixation or passage of the enzymes that exploit information, the polymerases. These are in fact enzymes of the *nuclease* type that represent by far the most useful family of molecules and therefore are more frequently used in genetic engineering. They can either detach fragments at the end (*exonucleases*) or cleave fragments at various places in the molecule (*endonucleases*). These two types of nuclease already coexist in the prokaryotes and have the primary function of protecting the genome of the cell they produce by degrading all the foreign DNA molecules that are able to penetrate it. It is moreover from various bacterial colonies that the enzymes are extracted, purified, and produced commercially. The endonucleases called restriction enzymes belong to this second category of enzymes; they can cleave the DNA chain at highly precise places and their specificity is precisely linked to the local chaining of nucleotides. These recognition sites, which are also cleavage sites distributed along the DNA molecule, generally comprise 4 to 6 base pairs. Depending on the enzymes, the cleavage can be a *straight cut* (e.g., *Eco*RV) or a *staggered cut* (e.g., *Eco*RI) (Fig. 1.34). This second mode of cleavage is presented much more favourably for a possible ultimate *ligation* (Fig. 1.35). Ligation, which aims to reassociate the DNA fragments cleaved by the endonucleases, is ensured by enzymes called *ligases*.

Name of enzyme	Extract of:	Recognized sequence	Type of cut
1 *Eco*R1	*E. coli*	...G AATTC... ...CTTAA G...	5' end
2 *Taq*I	*Thermus aquaticus*	...T CGA... ...AGC T...	5' end
3 *Bam*HI	*Bacillus amyloliquefaciens*	...G GATCC... ...CCTAG G...	5' end
4 *Hha* I	*Haemophilus haemolyticus*	...GCG C... ...C GCG...	3' end
5 *Kpn* I	*Klebsiella pneumonia*	...GGTAC C... ...C CATGG...	5' end
6 *Pst* I	*Providencia stuarti*	...CTGCA G ...GA CGTC...	3' end
7 *Dde* I	*Desulfovibrio desulfuricans*	...C TNAG... ...GANT C...	N = purine or pyrimidine
8 *Mst* II	*Microcoleus* sp.	...CC TNAGG... ...GGANT CC...	N = purine or pyrimidine
9 *Eco*RV	*E. coli*	...GAT ATC... ...CTA TAG...	Straight cut
10 *Hae* III	*Haemophilus aegytius*	...GG CC... ...CC GG...	Straight cut
11 *Dra* I	*Deinococcus radiophilus*	...TTT AAA... ...AAA TTT...	Straight cut

Fig. 1.34. Some examples of restriction enzymes and the sequences recognized by them. The types of cut of the DNA molecule are indicated.

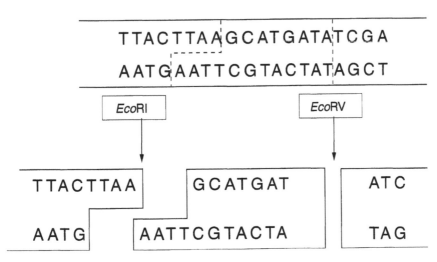

Fig. 1.35. Examples of cleavage by restriction enzymes. These cuts may be straight (*Eco*RV), i.e., at the same place in the two DNA chains, or staggered over two chains, generating "sticky" ends (*Eco*R1), i.e., always ready to hybridize. In the case of straight cuts, the ligation of the two fragments generated is sometimes difficult.

A certain number of experimental protocols have been fine-tuned for the search and identification of DNA fragments resulting from digestion by restriction enzymes. These protocols are based essentially on the mass of fragments. The rate of migration of these fragments on an electrophoresis gel is, in large part, inversely proportional to the mass and can consequently play a discriminating role. The problem is thus to observe on the gel the position of the fragment that has most often been amplified to cross the threshold of detection. This observation is generally made by means of a probe, radioactive or cold, according to the *Southern blot* technique, developed by M. Southern. Techniques based on comparable technology have been proposed for the gel detection of RNA (western blot) and proteins (northern blot). The successive application of these three tests suggests that isolated DNA clearly contains a functional gene.

b) DNA banks

Banks of genomic DNA are constructed by specific cleavages by a limited combination of restricted enzymes associated with ligation processes. This starts with the fragmentation of an entire genome of an organism, a plant, for example, by the same enzymes that serve to fragment plasmidic DNA of bacteria. The two DNA of different origins are then combined and the presence of ligases thus introduced in the medium allow the DNA to recombine, often with the risk of overlapping compatible ends. This allows in a certain number of favourable cases reconstruction of circular and functional plasmids made up of a DNA fragment from a genome of the plant and sufficiently large to contain the entire gene. These recombined plasmids are then put in the presence of bacteria (*E. coli*, for example), which, treated appropriately, tend to recover them. This recovery is effective to the extent that a selection pressure, to which the plasmid allows a response, is exerted on the bacteria (presence of an antibiotic in the culture medium and presence of a resistance gene on the plasmid). The bacteria thus reprogrammed can multiply actively and, consequently, amplify significantly the plant DNA fragment that the bacterial cells derived from this stock cell contain. Since each bacterium could have the same behaviour, the entire plant genome is found to be thus amplified. This important bacterial colony thus constitutes the *genomic DNA bank of the plant* (Fig. 1.36). It can be contained in a petri dish and easily distributed to laboratories that are interested in it. The bank contains the genes in their complete form, i.e., with introns and exons. However, some genes of the plant may not be found there if a cleavage by restriction enzymes has divided the gene sequence, which then ends up in two different fragments.

A second type of genomic bank exists: the *copy DNA or cDNA bank*. It presents some advantages over the preceding but also some disadvantages; the possession of two types of banks for a single species is

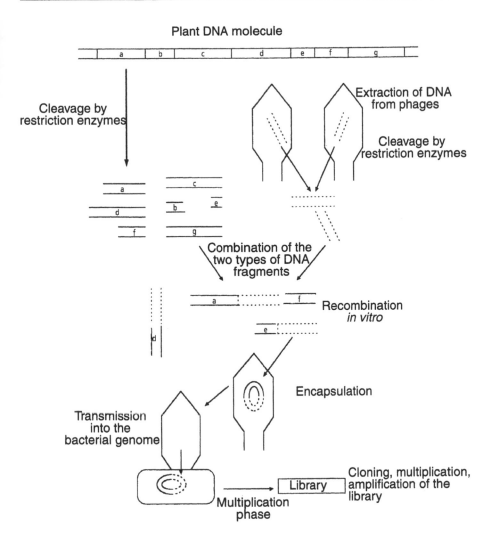

Fig. 1.36. Simplified diagram of the construction of a genomic library. The integrality of the nuclear DNA of developing young plantlets is extracted and cleaved by a restriction enzyme in a staggered cut. DNAs of phages carrying a character of selection (resistance to an antibiotic) are treated in the same way. The combination of the two DNAs, in the presence of ligases, leads to the formation of recombined DNA fragments containing the DNA fragments from the plantlets. These DNAs are recovered by phages that will infect bacteria. The bacteria multiply, form a significant colony, and constitute the "genomic library ", which can then be preserved in the frozen state. Subsequently, these plasmids can be extracted again and the plasmidic DNA used to transform mutant or disrupted yeast that may revert to a wild phenotype in case of "complementation of the deficient function".

not always superfluous. This type of bank is constructed using a particular enzyme, reverse transcriptase, which is capable of synthesizing DNA from messenger RNA. Also, generally it is started from a stage of development of the plant during which the maximum number of genes are found in transcriptional activity. An advanced embryo or very young plantlets are suitable but one can also consider a combination of messenger RNA extracted from different stages of development. These purified RNAs would serve as matrices for the synthesis of DNA, which would have the peculiarity of containing, in principle, only the minimal information, i.e., the reading frame of the genes without the introns. These DNA, placed under promoters, could give new messenger RNA that will have no need for maturation since they contain only combined exons. In reality, very often, upstream of reading frames, various sequences are found (copies of part of the promoter or control sequences) and the use of such banks demands a critical frame of mind. The ideal is to be able to have two types of "genes" and to compare their expression (Fig. 1.37).

This task of compiling banks is time-consuming and thus costly. Also, the scientific community has been in the habit of exchanging them on a more or less reciprocal basis. This has most often been the best and most economical means of procuring them. Some genetic engineering companies also construct and sell gene banks.

However, the task is far from being completed with the compilation of the bank. In order to construct genes voluntarily, the bacterium or bacteria that contain the "good" plasmid must be located in the bank, i.e., the bacterium that contains the desired gene reading in its sequence. This is the beginning of a long process of *screening*, which consists of finding the desired sequence by the most suitable method. There are several methods of achieving this. If a probe is used, i.e., a DNA fragment composed of a sequence of a few tens or hundreds of nucleotides complementary to the DNA to be identified, the method will consist of finding the DNA fragments capable of hybridizing by complementarity with this probe. This probe could easily be found by its emission of radiation if it is "hot", i.e., marked by an isotope (tritium, carbon 14, or phosphorus 32), or by a reaction of fluorescence or even colorimetry if the probe is "cold". When there is no probe available, there are other methods such as the detection of a protein coded by the desired nucleotide sequence. Still, in this case, the vector chosen to construct the bank allows the expression of sequences inserted.

Another effective technique is *functional complementation*. For research on plant genes, it is generally baker's yeast, *Saccharomyces cerevisiae*, that is used because it is probably the easiest and most effective material and contains the largest number of mutants. The principle consists of re-establishing a function in a mutant yeast that is

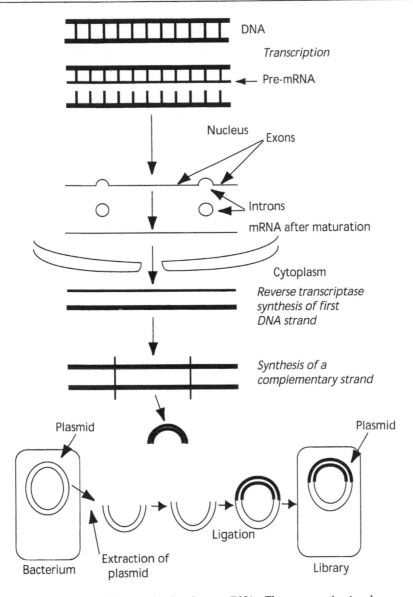

Fig. 1.37. Building a bank of copy DNA. The process begins by extraction of messenger RNA from young plantlets. In the presence of reverse transcriptase, these RNAs generate complementary DNA molecules, primarily single-stranded, which then duplicate. These DNA molecules are then cleaved by restriction enzymes and are combined, in the presence of ligases, with plasmidic DNA from bacteria cleaved by the same enzymes. As with the genomic library, the recombined plasmids are recovered by bacteria, colonies of which constitute the library of cDNA. These banks contain only the sequences coding for genes that have no introns.

deficient for that function (stage of biosynthesis, resistance or sensitivity to a product), by introducing plasmids from a bank of expression, one of which is capable of containing the gene of the corresponding plant. The "complementary" mutant yeast thus reverts to the wild type. In a colony of mutant yeasts, survival can only be ensured by the addition of a product to the culture medium that the yeasts cannot synthesize. Transferred to a minimal medium, all the yeast cells quickly die. Only those that, by chance, have in the course of a transformation by DNA from the bank (e.g., by electroporation) received a plasmid containing the desired gene can survive on such a medium. It is thus easy, after a few days at 28°C, to isolate the colony by making a culture of mass sufficient to extract from it plasmidic DNA containing the desired gene, "cleaving" this plasmid by the same restriction enzymes that served to make up the bank, then isolating by electrophoresis the DNA fragment containing the gene. A sequencing of this fragment, followed by a statistical analysis of the sequence (search for TATAbox and CAATbox, for example) generally allows the promoter to be identified and the initiation and termination codons and polyadenylation sequences to be found. The sequence can be compared with some 9 to 10 million sequences known worldwide and deposited in gene banks (such as the EMBL bank), and generally the gene in question can be identified.

When there is no mutant available to correspond with the desired function, it is still possible to create a *gene disruption*. In this case, DNA sequences are randomly introduced in the yeast genome until one of the yeasts *loses function* following the "rupture" of the gene into two segments separated by the introduced sequence. Such a gene is no longer functional but is in a way "labelled". This yeast finds its wild phenotype only in case a second transformation by DNA from the bank re-establishes the function. We therefore find ourselves in the same situation as before. These "labels" artificially inserted in the yeast genome behave like transposons and mime their effects.

To know *all* the genes and their nucleotide sequences in an individual seems to be the ultimate solution towards which all the studies on gene identification tend. Is this simple curiosity or an objective of high priority for fundamental knowledge? Opinions are clearly divided and the task is large enough so that even the most enthusiastic advocates have had to limit their ambition to the sequencing of a limited number of genomes: some bacteria, including *E. coli* (1997), two *Ricketssia* and *Agrobacterium* (2001), the yeast *Saccharomyces* (1996), the protozoa *Toxoplasma gondii* (2000), the worm *Caenorhabditis elegans* (1998; 10,000 genes), the fly *Drosophila* (2000; 13,600 genes), *Arabidopsis* (2000; 25,000 genes), rice (2001; 32,000 genes), human genome (2002; 32,000 to more than 100,000 genes if alternative splicing is taken into account) and the Popular (2004;

46000 genes). However, it must be emphasized that the knowledge of all the sequences of a genome does not imply the identification and precise location of all the genes. Much work remains to be done. This study is nevertheless a particularly effective tool for the purpose.

c) The mapping of genomes

Alongside these few privileged genomes, the great mass of genomes of all the other living organisms will never be known in entirety and can only be studied in a partial and often indirect fashion. This is not, however, an insuperable obstacle to projects of screened analysis of these genomes because there are other useful approaches in molecular biology. These include chromosome walking, RFLP and RAPD techniques, and some other approaches the complementarities of which have allowed important breakthroughs in the knowledge and control of the genome.

Chromosome walking consists of isolating the clones that carry genomic DNA sequences that partly overlap (Fig. 1.34). A large part of the DNA of a chromosome can also be "read" in the 5'-3' direction (some tens of Kbp) using a series of probes, each coming from a part of the fragment detected by the preceding probe. However, the first probe must correspond to a known gene. This is found at the starting point of the walk, as far upstream as possible, the walk being directed downstream toward the 3' end (Fig. 1.38).

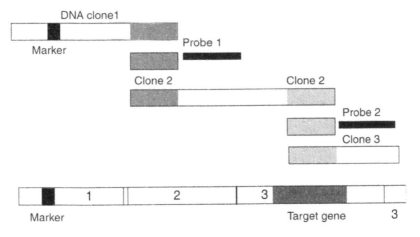

Fig. 1.38. Simplified diagram of the principle of "chromosome walking" to map the genome. This technique consists of using probes from 3' ends of relatively large DNA fragments cloned in a bank to look for regions overlapping different clones. These fragments are then reclassified to constitute the initial chain of these fragments in the initial DNA. If one of the fragments is labelled by a marker, the distance of a new gene from this marker can be specified.

In *Arabidopsis*, a certain number of genes have been discovered and published through the use of this technique. Examples are the gene ABI3, which controls the sensitivity of the plant to abscisic acid, AXR1, a gene for sensitivity to IAA, ETR1 or the gene for synthesis of ethylene receptor, and FAD3, which codes for synthesis of C18 : 2 desaturase.

The establishment of genome maps can also be based on RFLP techniques (restriction fragment length polymorphism), which is also called DNA restriction polymorphism (Fig. 1.39). This technique is based on the fact that between even two very similar individuals—such as two genotypes of the same plant variety—there are small variations that could cause a site of cleavage by a restriction enzyme to appear or disappear. We can also detect a difference in the size of fragments liberated when these DNA fragments are made to migrate on a gel and a probe is used to compare and locate precisely the bands corresponding to two individuals to be distinguished. As a preliminary approach, the technique primarily allows genotypic or varietal distinctions applied in agronomy to identify and authenticate claims of "property" in certain genomes and, if necessary, the payment of "royalties". It also allows sometimes the survival of the descendants of crosses carried out between two genotypes by following the mode of transmission of different restriction fragments within a single family, the modalities of possible recombination, and the measurement of genetic distances, as can be done with true genes. However, RFLP is a difficult technique because the probes chosen must be distributed quite homogeneously along the genome so as to constitute regularly spaced landmarks that are yet sufficiently close to recover lengths of DNA compatible with the size of recombination units of a genetic map.

Natural landmarks exist elsewhere in the genome also. These are the *DNA-satellites*, repeat sequences along the DNA. These sequences may be of varying lengths: some of them of average size (some hundreds to some thousands of bp); others much shorter (some hundreds of bp) constitute *minisatellites*. Still others, in which the motifs are very short and simple (often 2 bp repeated a good number of times), form the *microsatellites*. Variations between individuals pertain to the number of base motifs (TG, for example) and thus to the length occupied by this microsatellite on the DNA. Microsatellites are sufficiently widespread and regularly spaced along the DNA to be able to constitute reference marks. They can, with immediately adjacent DNA sequences, constitute primers that can be used for amplification reactions *in vitro*. The products of PCR are then analysed by migration on a polyacrylamide gel that allows the recording of variations of position due to differences in the length of the microsatellite. This technique is sometimes called *SCAR* (sequence characterized amplified region).

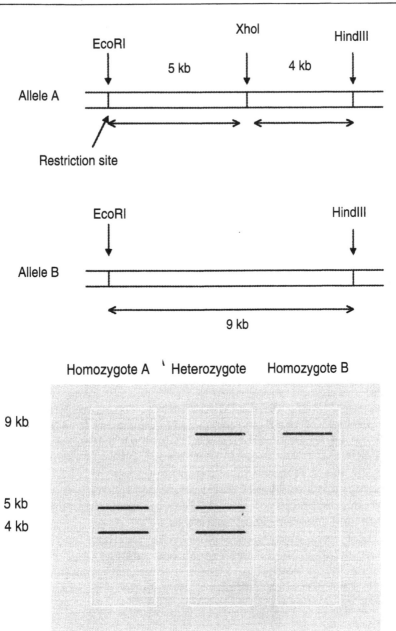

Fig. 1.39. RFLP technique. Two alleles of a single gene may not have the same sequence. If the difference pertains to the presence or absence of a characteristic sequence of a restriction enzyme, the fragments generated by an enzymatic digestion will not have the same length and that will result in two bands that do not have the same position on an electrophoregram.

The RAPD (random amplified polymorphic DNA) technique combines certain approaches of RFLP and PCR. It consists of amplifying the DNA fragments using the polymerization chain technique and using short sequences of primers in which the chaining of bases is purely conventional and random. The primers are very often polynucleotides of 10 to 20 bases, which will thus hybridize with genomic DNA at precise places at which these primers will find a complementary sequence. When the distance that separates these primers thus paired is compatible with the activity of the *Taq* polymerase used in the polymerization reaction (up to about 3000 bp), strands of DNA are synthesized and these, by their mass, faithfully record the distance. This could vary from one genotype to another and thus constitute a technique to distinguish, i.e., "identify", the genome of an individual.

There are still other methods based on the use of the polymerase chain reaction, associated with fragmentation of genomic DNA by restriction enzymes. The AFLP technique, for example, developed by the Keygene Society, recommends the ligation at two ends of digestion fragments of "bands" that will serve to fix specific primers, themselves responsible for the amplification of these fragments. This technique generates a very large number of bands and consequently increases the chances of discovering even slight differences in the structure of two genomes that are naturally less polymorphic. The reading of bands is, however, difficult and requires the use of integrators and appropriate software.

Research focusing on the identification of certain genes can use certain products of their expression as messengers. It is possible, for example, using reverse transcriptase, to go back to the copy DNA and consequently to a DNA sequence that has a role nearly identical to that of the gene itself. Genome labelling programmes by *EST (expressed sequence tag)* initiated by messengers have been particularly helpful for the cognitive approach of the *Arabidopsis* genome, particularly of redundant and non-redundant sequences.

All these techniques, used separately or together, lead to a deeper and more detailed understanding of the genome and constitute a prerequisite to the control of transfers from one genome to another, which is the particular objective of genetic engineering and its agronomic applications.

FURTHER READING

Plant biology and plant physiology

Buvat, R. 1989. *Ontogeny, Cell Differentiation, and Structure of Vascular Plants*. Springer Verlag, Berlin, Heidelberg, New York.

Coen, E.S., and Carpenter, R. 1993. The metamorphosis of flowers. *Plant Cell*, 5: 1175.

Heller, R., Esnault, R., and Lance, C. 2000. *Physiologie végétale*, 6th ed. Vol. S, *Nutrition*. Dunod, Paris.

Ridge, I. 2002. *Plants*. Oxford University Press, Oxford.

Rost, T.L., Barbour, M.G., Thornton, R.M., Weier, T.E., and Stocking, C.R. 1979. *Botany: A Brief Introduction to Plant Biology*. J. Wiley and Sons, New York, Chichester, Brisbane, Toronto.

Taiz, L., and Zeiger, E. 2002. *Plant Physiology*, 3rd ed. Sinauer Associates, Sunderland, Massachusetts.

Weigel, D., and Meyerowitz, E.M. 1994. The ABCs of floral homeotic genes. *Cell*, 78: 203-209.

Molecular biology and genetics

Albert, B., Bray, D., Lewis, J., Raff, M., Roberts, K., and Watson, J. 1994. *Molecular Biology of the Cell*, 3d ed. Garland Publishing Inc., New York.

Campbell, N.A. and Reece, J.B., 2005. Biology, 7th ed. Benjamin Cummings.

Chawla, H.S. 2002. *Introduction to Plant Biotechnology*, 2nd ed. Science Publishers, Inc., Enfield, New Hampshire.

Darnell, J., Lodish, H., and Baltimore, D. 1986. *Molecular Cell Biology*. Scientific American Books and Freeman and Company, New York.

Griffiths, A.J.F., Gelbart, W.M., Miller, J.H., and Lewontin, R.C. 1999. *Modern Genetic Analysis*. Freeman and Company, New York.

Lewin, B. 1995. *Genes V*. Oxford University Press, Oxford.

Maniatis, T., Frisch, E.F., and Sambrock, J. 1982. *Molecular Cloning: A Laboratory Manual*. Cold Springs Harbor Laboratory Press, New York.

Watson, J.D., Gilman, M., and Witkowski, J. 1997. *Recombinant DNA*. Freeman and Company, New York.

Widner, F., and Beffa, R. 1997. *Aide-Mémoire de Biochimie et de Biologie Moléculaire*. Lavoisier, Paris.

Chapter 2

Plant Biotechnology and Genetic Engineering

2.1. A NEW SCIENCE BUT AN OLD PARTNER OF AGRICULTURE

Agronomy is a complex science that emerged largely from the development of genetics associated with advances in physiology. It also owes a great deal to other major disciplines that have contributed to the rationalization of agriculture, including systematics, embryology, ecology, and pathology.

Physiology has been the source of major advances in cultivation practices, from the seed to the flowering plant. From genetics came large-scale programmes of plant improvement that particularly enhanced the value of the sequence from the flower to the seed. Agronomy thus brought a scientific dimension to agriculture that was essentially based on empiricism, pragmatism, prudence, and efficiency. As agronomy developed, it gradually transferred experimentation from the field to the laboratory and then applied the laboratory data to new field practices. Since the development of a plant occurs over time, all these techniques for the genetic improvement of plant material required a long time and the unit of time was often an annual *campaign*. This apparent "slowness" in the manipulation of plants contrasts with the rapidity of results obtained from the manipulation of microorganisms— bacteria and yeasts—in which the unit of time is mostly about one day and more rarely one week. In this difference of temporal scale we can see the origin of the high expectations that agronomists (or their sponsors) have invested in biotechnology in general and genetic engineering in particular, to shorten the time required for experimentation and thus to obtain a better return on their investment. The improvement of a cereal variety, for example, may take about ten years using conventional techniques of selection. Today that can be reduced to two or three years if the best biotechnologies are used. More than sixty years passed from the conception of the totipotency of the plant cell to its demonstration.

Less than ten years passed between the discovery of restriction enzymes and the first genetic transformations of plants. Only a few years passed before these plants reached our dinner tables, accompanied by a train of enthusiasts and protestors.

The intrusion of biotechnology into agronomy was neither sudden nor accidental but the result of a patient complicity affirmed every day and still far from having reached its potential. This complicity is still surrounded by secrecy and concerns that we must understand and respect. It is also marked by audacity and a wealth of projects that should be encouraged. And finally, it signals the dawn of a grand and probably rich adventure that we cannot and ought not to turn away from.

2.2. *IN VITRO* CULTURE

Genetic engineering in plants could not have developed so rapidly without a long prologue (of around three quarters of a century) of improvement, development, and cultivation of plant organs or tissues *in vitro*. These tissues and organs are, in most cases, essential to the regeneration of genetically transformed plants. They are indispensable to an operation of genetic transformation and so will be discussed at the very beginning of the chapter.

2.2.1. *In vitro* culture of organs and tissues

Plant biology was not the first field in which *in vitro* culture was attempted. The first assays were imitations of the experiments of Alexis Carrel on animal tissue culture in the early 20th century. According to the ideas of the time, culture was essentially the survival of these tissues in glass containers of varying forms. The tissues were kept alive by means of techniques of sterilization borrowed from Pasteurian microbiology, which was still new.

The precursor of *in vitro* culture in the plant world was Haberlandt at the very beginning of the 20th century. It was Haberlandt who stated clearly around 1902 his visionary idea of the *totipotency of the plant cell*, instantly generating high expectations about the future of this technology, which was then still in its infancy. The studies of the time show more a laboratory curiosity rather than a consistent scientific preoccupation. Those that followed would have been considered failures if the researchers had had any ambition other than to accumulate records of survival. Such research was set aside during the Great War and it was only in the 1920s, with Robbins, that it was taken up again, with little success. In the 1930s, the indefinite culture of root tips was recorded by White (1932), followed by the experiments of Gautheret at Paris and Nobécourt at Lyon. These latter studies took the subject to a higher level by considerably widening the field of research to numerous species and a wide variety of tissues. It is true that in the meantime certain growth factors were identified, such as auxins and gibberellins, which led to the development of culture mediums that could be considered standard, relatively multipurpose, and effective. The publication of *Traité de Culture des Tissues* in 1939 by Gautheret spelled out the knowledge acquired at that time, but research was again suspended during World War II.

After the war, research on this subject was resumed in different directions. Some scientists pursued ideas continuous with preceding research and contributed to the establishment of parameters and rules of *in vitro* tissue culture; these were always fundamentally useful and can be found in the treatises published by Gautheret and his colleagues from 1948 to 1977. Others chose to focus on increasingly smaller tissue fragments in the hope of reaching, or at least approaching, the culture of isolated cells. Still other researchers aimed to exploit these studies to identify possible applications in the field of agronomy. These orientations will be studied in the next section.

The chief impact of *in vitro* tissue cultures was *fundamental*: the acquisition of new and precise information about plant nutrition, the development of increasingly effective culture mediums, and the selective extractions that plants draw from the culture medium as well as their waste products. New hormones were discovered—cytokinins by Skoog in 1955 and the role of the auxin-cytokinin equilibrium in development processes—as well as the rational exploitation of properties of vegetative propagation naturally present in the plants from which tissue cultures were realized. Today it is probably the *applications* that keep the technology current. Tissue cultures are used to test the toxicity of certain products, their migration characteristics (apoplastic or symplastic), and the modalities of their detoxification. For example, many mitogenic properties (activators of division), mitostatic properties (inhibitors of division), or mitoclasic properties (destroyers of cells in mitosis) of products obtained by extraction or by synthesis have been and are now studied or tested on root tip cultures. Another common application of tissue culture is the production of metabolites derived from primary or secondary metabolism: production of perfumes and aromas (limonene, menthol, geraniol, anethol, etc.), dyes (carotenoids), or alkaloids with therapeutic effects (colchicines, vinblastin and vincristine, etc.). These applications fall within the fields of agronomy, agro-food, or bioindustry, depending on the quantities produced or the commercial value of the products.

Tissue cultures are initiated from a fragment of an organ (for example, a piece of stem, root, leaf, or tuber) the structure of which is as homogeneous as possible. The *explant* is taken by means of a scalpel, sterilized if necessary with diluted sodium (or calcium) hypochlorite, rinsed several times in sterile water, and cultured in a test tube or petri dish containing a sterilized culture medium. The culture medium is sterilized in an autoclave and sometimes, if there is a heat-labile fraction (some sugars, hormones, vitamins), by ultrafiltration. Nowadays, manipulation and transfer of the explant from one medium to another are carried out under a *laminar flow hood*, which considerably reduces the risk of contamination. The petri dishes or test tubes containing explants are kept in a *culture chamber* that has illumination suitable to the nature of the explant and in which as many parameters as possible are controlled (temperature, humidity, illumination with day/night alternation).

There are a great variety of culture mediums, if possible adapted to the genotype of the explant and to the research objectives. Even for the same plant, culture mediums differ depending on whether a simple increase in biomass or root differentiation is required. The mediums vary from one species or genotype to another in an unpredictable fashion. It does not seem possible to find a single "logic" to these variations, and culture mediums are still often developed according to empirical approaches. Despite all these difficulties, some specialists in plant nutrition have proposed standard mediums that will most often give correct results. Of these, the best known are the mediums designed by Kao, Murashige and Skoog, Gamborg, and Nitsch. Each medium has three parts: *macroelements* in large quantities, *microelements* in much smaller quantities, and variable amounts of *organic supplements* such as hormones and vitamins. All these products are generally dissolved in distilled water. The medium is often made into a gel by addition of agarose or agar-agar at a dose of less than 1% (Fig. 2.1).

In such culture conditions, the explant cells proliferate and the first visible result of that activity is the formation of a *callus*, an unorganized mass of cells that stick more or less together. In a *compact callus* the cells are very sticky and in a *friable callus* they are very loosely connected to one another and detach easily and often spontaneously at the edges. Such calluses can be used for the production of cell suspensions. Within the calluses, cell differentiation is generally poor but could involve the formation of conducting tissues. However, these tissue cultures sometimes lead to the formation of whole organisms, as has been done by Wetmore and subsequently by Wetmore and Morel on a fern (*Adiantum*). The photosynthetic capacity of cells is highly variable, often low, and a carbon source is nearly always added. The culture medium must periodically be renewed. Calluses can be

MICROELEMENTS			MACROELEMENTS			VITAMINS		
$CoCl_2$, $6H_2O$	0.025 mg/l	0.11 μM	$CaCl_2$	332.02 mg/l	2.99 μM	Glycine	2.00 mg/l	26.64 μM
$CuSO_4$, $5H_2O$	0.025 mg/l	0.10 μM	KH_2PO_4	170.0 mg/l	1.25 μM	Myo-inositol	100.00 mg/l	0.56 μM
FeNa EDTA	36.70 mg/l	0.10 μM	KNO_3	1900.00 mg/l	18.79 μM	Nicotinic acid	0.50 mg/l	4.06 μM
H_3BO_3	6.20 mg/l	0.10 μM	$MgSO_4$	180.54 mg/l	1.50 μM	Pyridoxin HCl	0.50 mg/l	2.43 μM
KI	0.83 mg/l	5.00 μM	NH_4NO_3	1650.00 mg/l	20.61 μM	Thiamine HCl	0.10 mg/l	0.30 μM
$MnSO_4$, H_2O	16.9 mg/l	0.10 μM						
Na_2MoO_4, $2H_2O$	0.25 mg/l	1.03 μM						
$ZnSO_4$, $7H_2O$	8.60 mg/l	29.91 μM						
	69.53 mg/l			4232.56 mg/l			103.10 mg/l	

Fig. 2.1. Basic composition of Murashige and Skoog medium, which is by far the most widely used for organ and tissue culture. There are many variants of this medium adapted to tissue types or particular species. Phytohormones in variable proportions are frequently added to this base medium.

fragmented and placed in a new medium, in a practically indefinite manner. These are successive *subcultures*.

2.2.2. *In vitro* culture of meristems

Meristematic cells constitute a particular "tissue" that merits separate consideration. This process is primarily carried out in the context of studies directed at reducing the size of the explant as far as possible without compromising its proliferative capacity. Shoot meristems and, to a lesser degree, root meristems are used for this purpose. The culture of this type of explant has, in fact, led to two new and original applications in the field of *plant protection* and that of *micropropagation* or *in vitro cloning* of plants.

a) Phytosanitary applications

In the field of plant protection, meristems are cultured in order to regenerate virus-free plants from an individual that has a virus. The method exploits a property revealed by the studies of two French researchers, Limasset and Cornuet, who demonstrated the heterogeneity of the distribution of viruses in plants and their near-absence in shoot meristematic cells. The excision of meristematic tissue and culture of these cells in mediums favouring processes of *regeneration*, i.e., the development of one or several individuals from the cell mass, result in individuals that are totally free of the virus. They are totally free, but not immune against fresh attacks.

Morel and Martin, in the early 1950s, succeeded in the regeneration of virus-free dahlias and potatoes. Their studies became famous because they involved the potato variety "Belle de Fontenay", highly valued before World War II and spoken of with nostalgia because it had reached a pathetic state following an accumulation of several viral diseases. The Belle de Fontenay varieties presently consumed, for the most part, are descendants of these *in vitro* survivors. This culture technique has fortunately replaced the techniques of *heat therapy* earlier applied with less success. Since reinfection by viruses is always possible, producers regularly renew their stocks, which can be "certified" as virus-free only if they meet precise specifications fixed by law.

Since these first studies, the production of plants from meristem cultures has developed and diversified considerably. Asparagus, strawberry, some varieties of rose, and several other plants that have industrial, horticultural, or market value have been improved and propagated using this technology.

b) "Micropropagation"

A second application, also interesting, arose from meristem cultures: *in vitro micropropagation*. A plant regenerated *in vitro* can be fragmented and then cultured further (as long as there is at least one axillary bud, i.e., the presence of a shoot meristem) to yield as many new individuals as there are fragments. These can in turn follow the same process and lead to the growth of a virtual population of individuals that are all identical, i.e., *clones*. When the results are compared with those obtained *ex vitro* by vegetative propagation, the greater efficiency of micropropagation can be observed. Many laboratories have used the technique to propagate individuals for experimental or commercial purposes (e.g., ferns, potato, tomato).

This technique was very early on applied to *in vitro* propagation of orchids (Fig. 2.2). The meristem taken from the embryo in the *protocorm* stage is fragmented in rigorously aseptic conditions and then cultured. Each fragment develops in time into a complete new protocorm, a regenerated one, from which the meristem can again be fragmented and so on. The use of cytokinins amplifies and accelerates this phenomenon. The development of orchid plantlets from these protocorms necessitates the complicity of a symbiotic fungus that is introduced in the culture medium and invades the cells of the lower part of the protocorm. This is called *artificial mycorrhization* and reproduces the events that occur naturally in the soil. Mycorrhization is far from being general but is found also in some ligneous species.

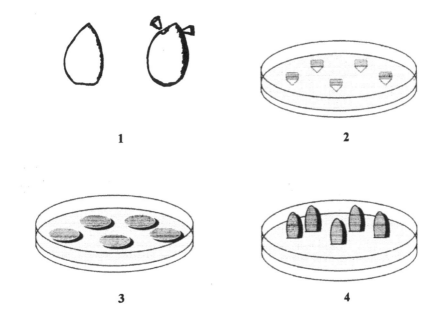

Fig. 2.2. Micropropagation of orchid from meristem fragments taken from the meristematic zone of a protocorm (1) cultured *in vitro* on MS medium (2). Each fragment rapidly gives rise to a small cell mass (3) that develops into new protocorms (4).

The technique of Micropropagation was also applied to strawberry, beet, cauliflower, raspberry, carnation, and other plants. In carnation, it is estimated that one meristem can be used to generate several millions of individuals in a few months. Micropropagation can be a thousand times as effective as simple natural propagation. This is industrial-scale production, but we must also take into account an inevitable mortality rate when the *in vitro* plants are *acclimatized*, i.e., transferred from the *in vitro* medium, despite the observation of precautions with respect to environmental conditions (saturating humidity, regulated temperature, scrupulously observed phytosanitary conditions). Very often, only these "acclimatized" plants can handle the usual culture conditions for the variety under consideration.

c) Cellular development characteristics

On the cellular scale, micropropagation involves an *organogenesis* that must be complete in order to be successful, i.e., it must involve *caulogenesis* or development of stem and branches, *phyllogenesis* or leaf

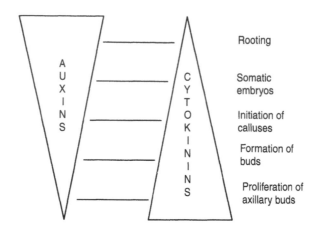

Fig. 2.3. Equilibrium between auxins and cytokinins and the corresponding characteristics with respect to the organogenesis involved.

formation, and finally *rhizogenesis* or development of the root system. Very often, various mediums must be used successively during the growth of the explant in order to ensure the sequence and progress of these ontogenetic programmes (Fig. 2.3). Root formation is often the most delicate phase and is found to be the main cause of failures. Correct rooting often requires a shift in the auxin-cytokinin equilibrium in favour of auxins; on the other hand, there are mediums that favour bud formation (low ratio of auxins to cytokinins). Also, sometimes the roots form but the vascular connections between the stem and roots are poorly established. For some cultures on the industrial scale, rhizogenesis is deliberately varied towards the period of acclimatization and the technology is also associated with a "cutting" of *in vitro* plants.

Despite the advances in our understanding of almost all the parameters that determine the success of these cultures, we must recognize that there are still many difficulties at all stages of cloning, often linked to the genotype or species concerned. This is particularly true of micropropagation of ligneous species. Ligneous species often require *in vitro* mycorrhization. Therefore, the procedures and methodologies must be adapted accordingly each time, which complicates the experiment and adds to its cost. This sector is still not highly mechanized, and abundant manpower and competence are required. Many small enterprises that embarked on the adventure of micropropagation have undergone great difficulties or have even sunk. Only sectors or departments associated with large industrial groups have been able to maintain this activity locally. There is some concern that the profitability of this sector should not necessitate a relocation of its activities to countries in which manpower cost is much lower and the legislation less restrictive.

In favourable cases that lead to the formation of somatic embryos, the embryonic processes are soon manifest in the formation of small nodules of meristematic cells (Fig. 2.4) that separate themselves from the surrounding cells by gradual rupture of the plasmodesmata connecting them to those cells, while the surrounding cells continue to be connected among themselves. This apparent structural separation is thought to correspond to a rupture of the trophic and hormonal correlations. The separation of cells within a mass is reminiscent of the situation of female gametes, then of the zygote and finally young zygotic embryos in the plant kingdom. Only cells belonging to these nodules are involved in the elaboration of somatic embryos, quickly achieving a polarized structure. Thus, there are the successive globular, heart, and "torpedo" stages of zygotic embryos. The difference lies most often in the absence of the *suspensor* or *haustorium*, which normally serves as an "umbilical cord" in the plant embryo but cannot usefully be cultured in a nutritive medium.

For a long time, scientists have investigated the conditions required and the mechanisms at work for one cell, among numerous others, to begin a process of regeneration. Is it the first cell to reach a threshold

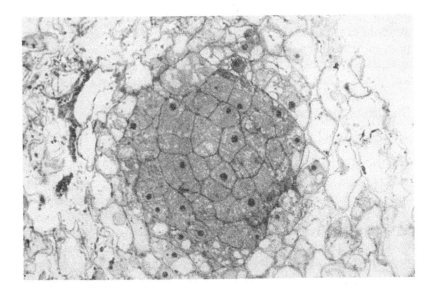

Fig. 2.4. Nodule of soybean cell, of meristematic type, involved in a process of regeneration within a tissue cultured *in vitro* (photo by D. Fournier).

stage of dedifferentiation or the first to succeed in withdrawing itself from the influence of its neighbours?

There is still no clear answer to this question, despite the large number of laboratories in which the process of regeneration is the focus of research. Some of the studies that we have carried out (Fournier, Lejeune and Tourte) suggest that a cell, to initiate an organism, must undergo structural and infrastructural modifications that are very similar to those undergone by an oosphere in the course of differentiation. The maturation of these gametes is often marked, apart from its isolation following the rupture of plasmodesmata and the secretion of new products, by the acquisition of highly sinuous curves of the nuclear envelope, the fragmentation of nucleoli, the formation of many micronucleoli, and the significant decondensation of chromatin. These nuclear events are accompanied by the densification of the matrix of mitochondria to the extent that they take on the shape of warped discs and the extreme simplification of the internal membrane structure of plastids. The appearance of particular intranuclear structures may evoke the formation of synaptonemal complex characteristic of the process of meiosis. Thus, it seems that to play the role normally assigned to the female gamete, the regenerating cell must repeat, in a highly telescopic manner, the gametophytic journey marked at its two extremities by sporogenesis and gametogenesis. An Italian researcher, Nutti Ronchi, has even described the existence of a virtual meiosis in the root cells of carrot. Similar observations were made in *Arabidopsis thaliana* in 2001. Do these observations not suggest the embryo development of mammals, which, in passing through a stage of formation of transitory and non-functional gills, briefly retrace their past as aquatic species? There is a principle often recalled in different biological groups that "ontogenesis retraces phylogenesis".

2.2.3. Culture of plant cells

a) From Haberlandt to Steward

The culture of isolated plant cells and the revelation of their potentialities are recorded in the general history of the progressive reduction of explant size and constitute the essential driving force of the development of plant technologies. The prophetic works of Haberlandt in 1902 on the totipotency of plant cells have haunted generations of researchers and stamped their imprint on the studies of Steward and his colleagues in the United States. These authors announced, in the late 1950s and early 1960s, that they had successfully regenerated a carrot from isolated root cells, thus realizing their dreams but also facing, at the time, incredulity from some French researchers who maintained the priority of their research on tissue cultures.

The progress from the isolated cell to the plantlet often passes through stages in which it is possible to recognize embryonic forms, especially with the presence of "cotyledons" and a gemmule-root axis. Thus, we speak of *somatic embryogenesis* even if this form of regeneration implicates none of the structures involved in the usual creation of an embryo by sexual means. This pseudo-embryonic development is successively involved in phenomena of cell division (meresis) and individual growth of cells (auxesis), as in the usual development of the embryo. These events, however, very often require the input of exogenous hormones, sometimes at high doses. The appearance of embryonic forms may be preceded by the formation of a callus called the *embryogenic callus*. It is from this callus that multiple embryonic structures form, which nevertheless are perfectly functional. This is called *indirect embryogenesis*. These "embryos" sometimes have anomalies such as the union of "cotyledons" in a kind of funnel at the base of which the "shoot" meristem cannot position itself or function (Fig. 2.5). Such "embryos" also often have difficulty in differentiating roots, which seriously compromises their ultimate development.

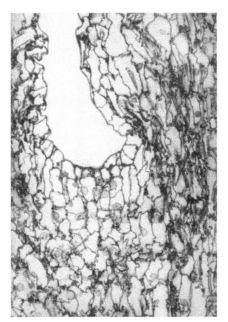

Fig. 2.5. Pseudo-embryo structures of soybean obtained by regeneration in *in vitro* tissue culture (left). These structures only rarely continue to develop in the absence of the establishment of a functional meristem (right) (photo by C. Bonnelle).

There are several methods for obtaining cultures of isolated cells. The first consists of using *friable calluses* derived from certain tissue cultures in which the peripheral cells easily detach themselves in an agitated medium. Cultured in the presence of auxins, these cells remain in suspension in densities compatible with the maintenance of their proliferative capacity. The cell populations are defective in not being highly homogeneous because of either their tissue origin or their degree of ploidy. Other techniques have been developed to reduce this disadvantage by starting with homogeneous tissues such as the palissadic leaf parenchyma.

Cells can be separated by so-called mechanical methods (razor blades) or enzymatic methods (pectinases). Apart from the base elements, the culture medium often has 2,4-D, NH_4^+ and K^+ ions, sulphates, and chlorides. Good results have also been obtained by *conditioning* the medium, either by a previous culture of an explant in the medium or by concomitant culture in two compartments separated by a filter. In one is a *nutrient*, a small, sticky cell mass, and in the second are isolated cells (Fig. 2.6). The capacity of cellular proliferation must not be confused with the embryogenic capacity of cells. The latter is not at its optimum from the time of culturing but develops after several weeks. It declines subsequently and tends to disappear in old cultures (Fig. 2.7). This loss of embryogenic capacity seems linked to an increase in the sensitivity of cells to auxin as well as a gradual increase in their degree of ploidy. The embryogenic capacity of a cell seems to be

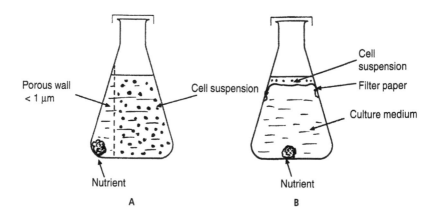

Fig. 2.6. Experimental set-up for proliferation of isolated cells in culture by the presence of "nutrients", small cell masses from a prior culture and separated from cells being cultured by a porous wall (A) or by a filter paper (B).

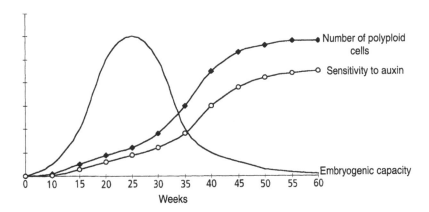

Fig. 2.7. Curves representing the loss of embryogenic capacity from the 25th week of culture. This loss seems linked to the increase in the number of polyploid cells and the acquisition of hypersensitivity to auxin.

associated with its capacity to conserve its diploid origin. However, it has been observed that the calluses resulting from cell cultures, which are incapable of evolving, are actually mosaics of different ploidies. In a tomato cell culture, for example, diploid cells may represent 25% of the population in the first days of culture and only 4% a few weeks later, while in the same time the rate of octoploid cells goes from 27 to 36% of the population. There are thus cellular derivatives during the aging of the culture, the ploidy perhaps being only a marker of this aging.

b) Artificial seeds

The technique of somatic embryo development seems well developed but only for a still limited number of species. The carrot is undoubtedly the best example of this and the production of somatic embryos today has reached the industrial scale. Asparagus, lucerne, some grasses, and horticultural plants are also produced. The results are still in their infancy for woody species apart from the date palm and banana. Research is being done on hevea, coffee, grapevine, and some conifers. For other species, all the stages are not yet mastered. These techniques can be improved for the development of *artificial seeds*. Somatic embryos raised in aerated liquid culture in a bioreactor, a kind of "fermentor", can be synchronized and calibrated by selection and then enclosed, or *encapsulated*, in a small alginate marble. This alginate compound, extracted from certain red algae, forms a gel in the presence of calcium, encloses the embryo, protects it, and thus forms a "seed" that can at any time be calibrated for the purpose of mechanical seeding. The alginate can also be incorporated in products favouring the ultimate development of the embryo, such as hormones, or in products protecting the embryo from bacterial or fungal diseases, as is done with real seeds.

c) The practical applications of isolated cell culture

Techniques of plant regeneration from isolated cells have largely been used in plant breeding because they make it possible to control a large population of individuals without the constraints of time and space that apply to the culture of whole plants. It can be used to select varieties resistant to some diseases (maize resistant to *Helminthosporium*) or varieties that are super-producers of amino acids and consequently high-yielding (tobacco, carrot), or resistant to certain herbicides of the 2,4-D type and the salinity of some soils.

2.2.4. Protoplasts

Isolated cell culture was developed in the 1950s, and the 1960s were marked by the culture and exploitation of plant protoplasts. In 1974, at Gif-sur-Yvette near Paris, the first international congress on the development of this technology was held under the aegis of the CNRS.

A plant protoplast (not to be confused with a proplast, which represents the juvenile or regressed form of a plastid) is a plant cell from which most of the cell wall has artificially been removed. This removal may be mechanical or enzymatic and the plant cells thus finds itself in a situation quite similar to that of the animal cell, separated from the outer environment by its intact plasma membrane covered by a polysaccharide felt that links it to the disappeared wall and is comparable to a "cell coat". Under the pressure of vacuolar turgescence, the cell will tend to burst if its integrity is not maintained by an artifice in the composition of the culture medium: the introduction of a non-metabolizable or poorly metabolized sugar in such a way that its concentration hardly varies over time. The sugar most often used in this case is mannitol added to a culture medium at a dose of about 10 to 13%. Apart from this addition, the culture medium is often the same as those used for the culture of cells in their mineral as well as organic composition.

a) Obtaining protoplasts

Protoplasts are isolated at first by plasmolysis from the palissadic leaf parenchyma of tobacco followed by laceration using a razor blade. Plasmolysis results in the disappearance of all the plasmodesmata and the retraction of the cellular contents in a reduced volume of the original cellulose framework. The laceration of the tissue destroys a large number of cells and also opens a certain number of cell wall frameworks and liberates protoplasts through the breaches thus created (Fig. 2.8). As soon as they are liberated, the protoplasts take a spherical form and disperse in the medium. The dispersion is facilitated by the presence of electronegative charges on the outer surface of the protoplasts that tend to repel, as do magnets that have the same charge. This technique has a relatively low yield (number of protoplasts to number of cells treated).

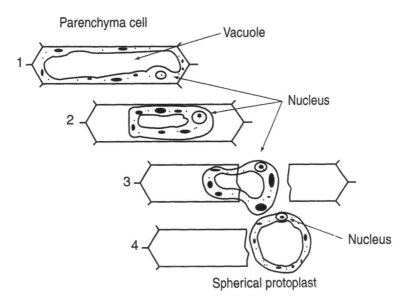

Fig. 2.8. Successive phases of formation of a plant cell protoplast after plasmolysis and mechanical rupture of the wall. The protoplast quickly becomes spherical in the culture medium, in which the concentration of some sugars must counteract the absence of the wall.

Researchers nowadays prefer to isolate protoplasts by enzymatic means. The digestion of cell walls involves pectolytic and cellulolytic enzymes: pectolytic enzymes attack the pectic cement, also called the middle lamella, which holds the cells together, and cellulolytic enzymes attack the wall itself. The first enzymatic formulas used came from the maceration of a combination of fungi known for their lignicolous properties. This enzymatic combination, called "macerase", contained only 10% of the active ingredient and many impurities, so the manipulations were not reproducible. The industry of fine chemistry has perfected the modes of extraction and purification: the enzymes proposed for this type of digestion have improved in quality (cellulase extracted from *Aspergillus*, driselase of fungus, helicase of snail) and the preparation of protoplasts has improved in reproducibility. Nowadays, the range of enzymatic preparations is quite rich, making it possible to manipulate protoplasts in university laboratories.

Segments of the leaf blade, with as many veins removed as possible, are immersed in a digestive medium placed in the dark at 20°C, for a duration varying from a few hours to overnight. The protoplasts are then centrifuged at very low speed (at around 100 g) so that they are separated from the leaf residues and then washed several times in a buffer medium containing a good quantity of mannitol. They are then cultured in petri dishes (at a ratio of around 50,000 protoplasts per ml) containing a liquid or gel culture medium. Their development can be followed under inverted light microscope. The first manifestation of activity is the formation of a wall, which can be made visible by staining with calcofluor in a 1% buffer solution. If the initial cell comes from a relatively young parenchyma, the number of sources from which the wall develops is high, while in the case of an aged leaf parenchyma, the number of sources is much smaller (Fig. 2.9). In any case, the new wall attaches itself to the spherical form of the protoplast and it is therefore easy to distinguish, in a culture, between the cells isolated from the original culture and protoplasts that have a reconstructed wall. Subsequent manifestations involve the resumption of mitotic activity with a frequency and rhythm characteristic of the species and leading to the formation of a callus or an embryonic form, depending on the case. The procedure for protoplast culture is thus identical to that for isolated cell culture. It is, however, possible to retain the protoplast character by maintaining the presence of cellulolytic enzymes in the environment.

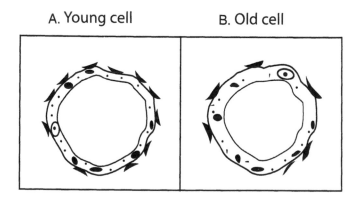

Fig. 2.9. From the time protoplasts are isolated, they begin to reconstruct their wall from secretion sources. These sources are numerous if the protoplast is derived from a young cell (A), but they are much fewer when they come from a differentiated cell (B).

b) The fusion of protoplasts

Protoplast culture became the focus of interest when technical advances made possible the *fusion of protoplasts followed by regeneration of fusion products*. For this, it was necessary to reduce and even annul the electrical charges present on the surface that keep protoplasts apart. Two fields of research have been explored in this regard: chemical methods and physical methods. A preliminary solution came from the use of polyethylene glycol or PEG, a water-soluble molecule with a molecular mass ranging from 1500 to 6000 that has the property of forming a film at the periphery of the protoplast and thus masking the electric charges. The protoplasts no longer find it difficult to approach one another and, in the most favourable cases, to fuse. This technology is now well developed but the dose of PEG required for fusion is very high (30% solution) and it is not easy to separate it from the culture.

Electrical methods relying on membrane depolarization have also been developed in parallel. High frequency currents applied directly in the culture medium by means of electrodes can be used to obtain chains of joined protoplasts parallel with the forces of the electrical field thus created. These currents applied for a few minutes are followed by an electric discharge of very short duration (a few milli- to microseconds) but high intensity (several hundreds of volts per cm). The electrical shock leads to the death of a large number of cells but at the same time induces numerous fusions between neighbouring protoplasts. It is believed that the electric discharge, in causing the displacement and combination of hydrophilous proteins in the lipidic bilayer, leads to the formation of virtual canals favourable to the removal of water and living matter as well as to the continuation of two plasma membranes belonging to two contiguous protoplasts (Fig. 2.10). This constitutes an essential and sufficient prerequisite for the fusion of two partners. These partners can have the same origin, i.e., come from the same plant or two plants belonging to the same species (homologous fusion) or from two different and even sometimes distant sources (heterologous fusion). In the latter case, the fusion of protoplasts from a parenchyma of a normal plant and from a parenchyma of a plant mutant for chlorophyll synthesis (albino) is easy to identify. The same is true when the two colonies do not have the same sensitivity to an antibiotic. But when the fusion is homologous, one of the major difficulties lies in separating the products of fusion from the initial protoplasts.

Analogous techniques can be used to cause the fusion of protoplasts and liposomes. Liposomes are artificial phospholipid structures formed of concentric layers in which the structure is fairly similar to that of biological membranes. These structures may fuse with membranes and liberate the products, including DNA, previously encapsulated in the

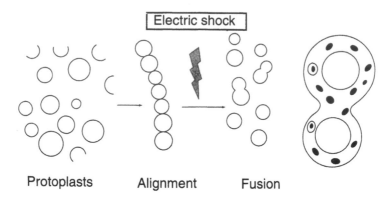

Fig. 2.10. Electrofusion of protoplasts. The use of a high frequency current generator favours the formation of small chains of protoplasts that come close enough to one another to touch and fuse when the electroporator sends a high electric current into the medium.

liposomes. We shall discuss these structures again when we study the technical modalities of genetic transformation.

Protoplasts have also successfully been used to absorb large biological structures such as bacteria, yeasts, cyanobacteria, and plastids.

c) Somatic hybrids and cybrids

The fusion of two protoplasts gives rise to a tetraploid cell (4n). The organism that thus results from regeneration, unless regulation processes intervene, is also tetraploid. It may be interesting to compare the organism resulting from sexual reproduction with that derived from a somatic fusion because, apart from this difference at the level of ploidy, the latter inherits two parental cytoplasms and the organelles they contain, while the former receives, in most cases, only the maternal cytoplasm. In the best cases, this structure is stable and conserves its two parental genomes and its hybrid cytoplasm.

However, in some cases it has been shown that in the course of ontogenic divisions one of the two genomes is progressively eliminated and the structure evolves towards an original situation that comprises only one of the genomes and the two cytoplasmic components. This is called a *cybrid*. The elimination of one of the genomes may be sudden and very early in the development phase. One example is a cross between two lucernes: *Medicago sativa* and *M. falcata* (2n = 32), one with yellow flowers and one with violet flowers, which gives a hybrid with combined flower colours. Researchers (Y. Dattée et al.) have obtained somatic hybrids that present the same phenotypic features as the hybrids, but at 4n = 64. However, in this case as well, there have been descendants that have lost either part or all of one of the genomes. Today there are numerous intermediates ranging from the somatic hybrid (4n = 64) to the strict cybrid (2n = 32). Even in this last case, both the cytoplasms are conserved. The comparative study of agronomic qualities of all these plants arising from the sexual pathway and the somatic pathway is of great interest in terms of genetics.

Theoretically, there is no limit to somatic hybridization. Intergeneric, interfamilial, and even inter-individual fusions have been realized in the animal and plant kingdom (e.g., *Xenopus*/carrot or tobacco/HeLa cell) but have never resulted in an organism.

d) Practical use of protoplast culture

Protoplast technology has many applications. It would be futile to attempt to cite all the sectors of research that have benefited from it. Some examples, besides somatic hybridization, are studies undertaken on modalities of the cell division cycle, secretion of plant metabolites, mechanisms of cell differentiation, characteristics of membrane transport, resistance to viruses, bacteria, and fungi, somaclonal variability, and selection of mutants.

Somatic hybrids and cybrids, as we have just seen, are of great interest in fundamental research. There are also numerous applications that have an impact on the evolution of our understanding and on agronomic practices. Unfortunately, all plants do not respond favourably to protoplast technology and especially to the regeneration of fusion products. Solanaceae (e.g., tobacco, tomato, potato, petunia) are good candidates. Good results have also been obtained with Apiaceae (carrot, fennel) as well as with Brassicaceae (rape, cress, radish, wild radish). Other families (Fabaceae, Poaceae, Asteraceae) are considered recalcitrant even though some members have sometimes given positive responses, such as soybean, rice, maize, barley, and sunflower. Among the plants belonging to a single species, it has been easier to get successful results with varieties that are still close to the wild types. Generally, in plants that have undergone selection and improvement, there is greater difficulty in regenerating from protoplasts.

Not the least of the advantages of the somatic route is the possibility of overcoming the barriers of speciation that generally prevent extraspecific hybridization. The most famous example is the creation of the *pomato*: a somatic hybrid of potato and tomato that was more of a theoretical advance than an agronomic success. There have been many other attempts (*Arabidopsis*/*Brassica*, *Atropa*/*Petunia*, *Nicotiana*/*Petunia*), but they have very rarely resulted in flowering and still more rarely in fertility.

2.2.5. *In vitro* culture of haploids

a) Principles

In plants, the haploid phase has an importance that it never has in animals. It goes from the spore to the gamete. This sequence is even more important in the lower Cormophytes, mosses and ferns, but till considerable in the higher Cormophytes (from the microspore to the sperm nucleus and from the macrospore to the oosphere). Agronomists have always been greatly tempted to induce the expression of the ontogenic potentialities in a cell derived from meiosis under perfectly controlled conditions.

In nature, some plants develop spontaneously from non-fertile reproductive cells, as we have seen in Chapter 1. Such instances are grouped under the term *parthenogenesis*. In their fundamental principle, these plants are haploid and sterile. In reality, processes of regulation very often re-establish their diploidy and their fertility. These plants are, moreover, totally homozygous. Researchers are therefore very interested in understanding and reconstructing the conditions in which such phenomena occur.

There are two ways of realizing parthenogenesis, the male route or *androgenesis* and the female route or *gynogenesis*. The male route is by far the most frequently studied.

b) Androgenesis

The cell resulting from meiosis, the microspore, is made to express its sporophytic programme in place of its gametophytic programme, which normally occurs at this stage (Fig. 2.11). *In vitro* culture of these microspores, creating an unusual environment, seems to be the easiest solution to implement. Practically, the first stage consists in establishing, for the species or variety chosen, a correspondence between the size of the flower bud and the precise stage of evolution of the microsporogenic meiosis. The microspore must end its second meiotic division, be separated from its neighbour, but not yet have begun its process of transformation into a pollen grain. This stage corresponds most often to a floral button that is still small and tightly closed but, when dissected under a microscope, shows stamens in which the different parts are clearly individualized but in which the clefts of dehiscence of the anther are not yet pierced. A slight pressure on the anthers precipitates this dehiscence and allows the microspores to be collected at a suitable stage. There are nevertheless many differences between species and the preliminary task of locating the stages of maturity must be adapted to each case.

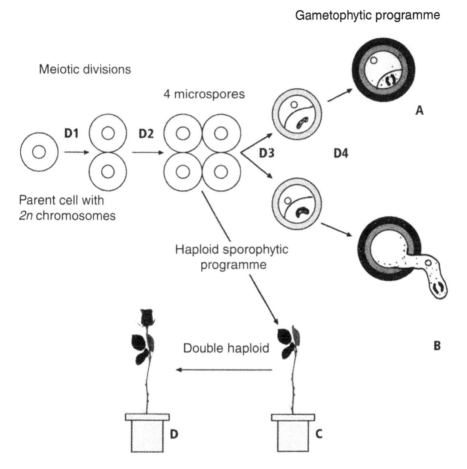

Fig. 2.11. Androgenetic development. Following the two divisions (D1 and D2) of meiosis, the parent cell yields four microspores with *n* chromosomes. These microspores develop normally, according to the gametophytic programme they carry, into bicellular pollen grains (D3) and, in some species, trinucleate pollen grains (D4) before even the germination of the tube (A). The last karyokinesis takes place, in most cases, only during the growth of the tube (B). In cases in which microspores are cultured *in vitro* in a suitable medium, the sporophytic programme, which they also carry, leads to the formation of sterile haploid plants (C). Spontaneously or artificially by means of colchicine, the genome could become diploid and, in this case, the plant is 100% homozygous. It thus becomes fertile (D).

GAMBORG MEDIUM

MICROELEMENTS			MACROELEMENTS			VITAMINS		
$CoCl_2$, $6H_2O$	0.025 mg/l	0.11 μM	NaH_2PO_4	130.40 mg/l	1.09 μM	Myo-inositol	100.00 mg/l	0.56 μM
$CuSO_4$, $5H_2O$	0.025 mg/l	0.10 μM	KNO_3	2500.00 mg/l	24.73 μM	Nicotinic acid	1.00 mg/l	8.12 μM
FeNa EDTA	36.70 mg/l	0.10 μM	$MgSO_4$	121.56 mg/l	0.51 μM	Pyridoxin HCl	1.00 mg/l	4.86 μM
H_3BO_3	3.00 mg/l	48.52 μM	$CaCl_2$	113.23 mg/l	1.01 μM	Thiamine HCl	10.00 mg/l	29.65 μM
$MnSO_4$, H_2O	10.00 mg/l	59.16 μM	$(NH_4)_2$ SO_4	1650.00 mg/l	134.00 μM			
Na_2MoO_4, $2H_2O$	0.25 mg/l	1.03 μM						
$ZnSO_4$, $7H_2O$	2.00 mg/l	6.96 μM						
KI	0.75 mg/l	4.52 μM						
	52.75 mg/l			4518.20 mg/l			112.00 mg/l	

NLN MEDIUM

MICROELEMENTS			MACROELEMENTS			VITAMINS		
$CoCl_2$, $6H_2O$	0.025 mg/l	0.11 μM	KH_2PO_4	125.00 mg/l	1.23 μM	Glycine	2.00 mg/l	26.64 μM
$CuSO_4$, $5H_2O$	0.025 mg/l	0.10 μM	KNO_3	125.00 mg/l	0.92 μM	Myo-inositol	100.00 mg/l	0.56 μM
FeNa EDTA	36.70 mg/l	0.10 μM	$MgSO_4$	61.00 mg/l	0.51 μM	Nicotinic acid	5.00 mg/l	40.06 μM
H_3BO_3	10.00 mg/l	0.16 μM				Pyridoxin HCl	0.50 mg/l	2.43 μM
$MnSO_4$, H_2O	18.95 mg/l	0.11 μM				Thiamine HCl	0.50 mg/l	1.48 μM
Na_2MoO_4, $2H_2O$	0.25 mg/l	1.03 μM				D(+)-biotin	0.05 mg/l	0.21 μM
$ZnSO_4$, $7H_2O$	10.00 mg/l	34.78 μM				Folic acid	0.50 mg/l	1.13 μM
						L-glutamine	800.00 mg/l	5.47 μM
						L-serine	100.00 mg/l	0.95 μM
						Glutathion	30.00 mg/l	0.10 μM
	75.95 mg/l			311,00 mg/l			1038.55 mg/l	

Fig. 2.12. Composition of Gamborg and NLN mediums, which are used in microspore culture for orientation towards the execution of the sporophytic programme.

Once these indicators are taken, the second stage consists of conducting the same operation in rigorously aseptic conditions and on a certain quantity of buds taken at the same stage of development. The microspores are collected in Petri dishes containing a culture medium that must also be adapted to each species, starting from a standard medium. The medium must be relatively rich and contain a source of carbon, minerals, amino acids, vitamins, and hormones. The Gamborg medium is often the first to be used. It is soon replaced by Nitsch's NLN medium (Fig. 2.12). The culture dishes are then placed in the dark in an oven at 28°C for about 15 days and then in a culture chamber between 22 and 25°C in light in an agitated culture stage. After a week, the small embryos that have developed are transferred to a gel medium, first in Petri dishes, then in tubes, and cultured always in the same conditions. They are then taken out of the tubes before being acclimatized for greenhouse culture. Regular controls under inverted microscope are recommended. The first manifestation of the establishment of a sporophytic programme expected is the symmetry of the first mitosis while the gametophytic programme begins with a highly asymmetrical mitosis. Regular mitoses lead to a small subspherical cell mass, a sort of small "morula", which as it develops may take on an embryoid form reminiscent of the development of cotyledons laterally flanking an embryonic axis. The programme is then well in line and must result sooner or later in the development of a young plantlet. After successive subcultures in aseptic conditions and culture under illumination adapted to the species, the plantlets can be taken out of the tube and, after acclimatization, be raised in the greenhouse.

The plant is normally haploid but only a precise counting of the number of chromosomes can confirm this because its phenotype will not look different from the normal phenotype except to particularly trained persons. Generally, we must wait for the flowering of the plant and observe its sterility in order to confirm this. The plants that are seen at this stage to be fertile generally show, during the first divisions, one or several karyokineses not followed by cytokinesis, which have spontaneously re-established diploidy. Diploidy can also be re-established by treatment of the plantlet with a low-concentration colchicin solution (mitostatic dose, 1%) followed by subculture on a control medium. A certain fraction of the population regenerates, from treated shoot meristems, stems in which the chromosome complement of the cells is diploid. Cuttings from these stems result in fertile plants. These plants, with a genome strictly conforming to the genome of the microspore, are 100% homozygous and, consequently, have all the conditions required to ensure high stability across generations. They are *haplodiploid* plants and the operation is called *haplodiploidization*. The *double haploid* plant is thus achieved in one operation, whereas it would have required about ten generations of self-fertilization in a genetic

Acacia sp.	*Lilium longiflorum*
Beta vulgaris	*Linum usitatissimum*
Brassica napus	*Lolium perenne*
Brassica oleracea	*Medicago sativa*
Capsicum annuum	*Oryza sativa*
Coffea arabica	*Pinus sp.*
Daucus carota	*Solanum phureja*
Eucalyptus communis	*Solanum tuberosum*
Fagopyrum esculentum	*Triticum aestivum*
Freesia spp.	*Triticum durum*
Helianthus annuus	*Vitis vinifera*
Hordeum vulgare	*Zea mays*

Fig. 2.13. Examples of species, among the most common in agriculture, that have produced haploid plants from microspore culture.

improvement programme based on sexual fertilization. These two advantages of quick results and homozygous purity very quickly attracted the attention of agronomists, and many programmes of androgenesis have been developed, particularly in cereals and Brassicaceae. The technique will soon be about 30 years old and the list of successful experiments is very long (Fig. 2.13).

Variants of microspore culture have also been developed, all of them aiming to make the experimental protocol easier. Culture of entire anthers, reducing the "sampling" step, gives very good results in barley, rape, and other plants. Anthers extracted at a suitable stage are surface sterilized, directly cultured on a liquid or solid nutrient medium depending on the species, and cultured in conditions very similar to those required for microspore culture.

As with somatic cells that regenerate from an organism, there may be spontaneous variations due to a variability in the number or structure of chromosomes that could be a source of genetic variability. These variations, referred to as *gametoclonal variations*, are widely used in agronomy, especially in the improvement of cereal varieties.

c) Gynogenesis

The controlled development of an organism from a haploid cell belonging to the female line is much less advanced. The main reason for this is the inaccessibility of the female cell and its sister cells in the embryo sac. This difficulty has been overcome by culture of the organ that contains the female cells: the ovule and sometimes even the ovary. Here too, the culture must coincide with a precise stage of development of the female gametophyte. This stage is not always well identified and many failures can be accounted for by cultures that were too early or too late. Calluses that are more or less chlorophyllian may appear and, in a still small number of cases, give rise to a haploid plant. The first successes were reported in 1976 for barley, 1979 for tobacco, and 1980 for wheat and rice. Today there are about 20 species distributed over about 15 families that have been cultured in this way. There is no systematic "logic" and success very often depends on a high level of empiricism associated with a few technologies (pre-conditioning of plants, cold pre-treatment of floral buds, hormonal composition of culture mediums). The yield is always relatively low and even expectations based on the appearance of calluses are often disappointed because of a nucellar origin of tissues. The technique is used, however, for species that have not responded favourably to microspore culture (e.g., beetroot).

d) Advantages of haploids in agronomy

The obtaining and culture of haploids, principally double haploids, is a growing field in the sector of plant improvement. The technique works best for obtaining plants that are totally homozygous and therefore constitute *pure lines* that can be used in crosses for pedigree selection and thus faithfully follow Mendelian laws. These plants express all the characters present in their genotype, including characters that were recessive in a hybrid situation, and thus restore the haploid genetics so valuable to geneticists for genotype/phenotype conformity. Compared to an F_2 generation, this technique makes it possible to define a population more quickly for the purpose of selection or to reduce the size of the population from which to select while preserving a similar probability of success.

In practice, these techniques have been used to select characters of precocity, size, tolerance to *Phytophthora*, and level of alkaloids in tobacco, precocity, fertility, and protein content of grain in rice, and yield, size, resistance to parasites, and precocity in barley. They are adapted to autogamous as well as allogamous plants.

2.3. GENE TRANSFER IN PLANTS

2.3.1. A practice as "old" as life

After each sexual cross, large quantities of genes are transferred from one part of the stamen to the oosphere via the pollen tube. Each fertilization thus ends in the reconstruction of an original genome belonging earlier to the embryo from the addition of two parental genomic blocks. In this the plant simply imitates the bacteria that since the dawn of time have transferred genes to one another by a process called *conjugation*. They can also use the viruses that periodically infect them to exchange genetic information in a process of *transduction*. These exchanges occur not only between prokaryotes but also between eukaryotic cells, such as yeasts, via bacterial plasmids. It is known that, even within the cell, there are exchanges between the nuclear genome and the genomes of plastids and mitochondria. Finally, a study on the molecular scale of a fairly serious and common parasitic disease of certain Dicotyledons has demonstrated the occurrence of gene transfer between bacteria and plants, mechanisms that molecular biology exploits for a perfectly controlled transfer of genes from one plant to another.

2.3.2. Crown gall disease and natural genetic engineering

a) Manifestations of crown gall disease

The relationship between soil bacteria of the genus *Agrobacterium* and the occurrence of tumours in the crown, i.e., the zone between the root and the stem, in some Dicotyledons, particularly cabbages (*Brassica*), has been known for a long time. This tumour is generally invasive, leading to necrosis of the above-ground part of the plant and finally its death. An anatomical section of the tumour shows intensive multiplication of cells of all the tissues present at the crown, very closely linked to considerable bacterial proliferation. This bacterial proliferation is supported by the production, by the tumour cells, of *opines*, new molecules that are never present in a healthy plant. Opines serve as nutrients for agrobacteria and activate their multiplication but their synthesis hijacks some pathways of normal metabolism in the plant. About 20 years ago it was demonstrated that this deviation of the metabolism of plant cells results from a virtual operation of "natural genetic engineering" carried out by agrobacteria and the large plasmids of about 200 kbp that they harbour.

There are several strains of agrobacteria, the two best known species being *Agrobacterium tumefaciens*, which harbours the Ti (tumour-inducing) plasmid, and *A. rhizogenes* (two ≠ species), which harbours the Ri (root-inducing) plasmid and causes, in the parasitized plant, *hairy roots*, marked by an abnormal development of the root system in quantitative terms (numbers of roots) and qualitative terms (long roots). These are gram-negative bacteria belonging to the category Eubacteria.

b) Mapping the Ti plasmid

The large circular plasmids of double-stranded DNA are present in the form of several identical types. They comprise different regions, among which three are particularly important (Fig. 2.14):

• a region containing the origin of replication;
• a region corresponding to the transfer of DNA or T-DNA, limited by two borders known as left and right and containing oncogenes responsible for the tumour;

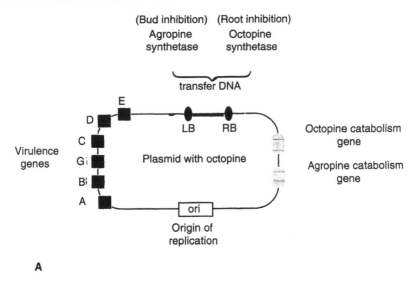

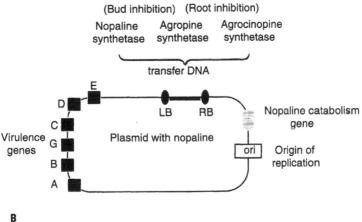

Fig. 2.14. Simplified map of Ti plasmids with octopine (A) and nopaline (B).

- a *vir* region containing various genes influencing the virulence of the bacterium, i.e., its capacity to recognize plant cells, infect them, direct the transfer of T-DNA into them as far as the nucleus, and integrate that T-DNA into the genome of the plant cell.

The integration of T-DNA into the genome of the plant cell leads to the modification of cellular metabolism and loss of the cell's control over mitotic orientation and mechanisms. This results in the formation of the tumour and the secretion of opines.

The opines are a family of molecules comprising generally an amino acid linked to a sugar. These two components may vary from one type to another, which has led biochemists to establish three classes of plasmids according to the nature of opines:

- plasmids with *octopine*, in which the sugar is pyruvate and the amino acid is arginine, ornithine, lysine, or histidine.

- plasmids with *agropine*, in which the amino acid is glutamine, generally associated with mannose;

- plasmids with *nopaline*, in which arginine or ornithine is found to be linked to acetoglutarate.

Plant cells that have been reprogrammed by bacteria to realize these syntheses are not capable of utilizing such molecules. In contrast, there are agrobacteria having genes located precisely on these Ti plasmids, whose products of translation are capable of degrading these molecules and utilizing the products of degradation as nutrients.

The T-DNA also contains genes that code for the synthesis of enzymes involved in the biosynthesis of hormones. The genes *iaaM* and *iaaH* code for enzymatic proteins that are involved in the synthesis of an auxin. The gene *iptZ* codes for the synthesis of a cytokinin; these two types of hormones intervene in cellular proliferation of the receiving plant. The T-DNA is flanked by conserved sequences of about 25 bp that constitute *borders* corresponding to excision zones and thus helping to liberate one of the strands. The cleavage occurs first at the *right border* and then at the *left border*, these terms being defined with respect to the origin of the replication located on a representation of a plasmid on a map. The transfer thus involves a single-stranded DNA that will behave rather like a mobile transposable element capable of integrating itself into the genome of the host plant. This transfer requires direct contact of the bacterium with a plant cell, which is like an original process of *conjugation* between a prokaryote cell and a eukaryote plant cell.

The *vir* genes, located on the plasmid but outside the T-DNA region, are essential to the transfer. There are at least eight operons located linearly on the plasmid, in a region neighbouring the origin of the replication and identified by letters of the alphabet as follows:

- *vir A* codes for the synthesis of a receptor of an injury signal, a receptor sensitive to molecules when there is accidental injury to a plant.

- *vir B* codes for the synthesis of a membrane protein intervening in the formation of a channel through which the T-DNA molecule must travel during its passage from the bacterium into the plant cell.

- *vir C* directs the synthesis of a protein capable of directing the transfer of the T-DNA, placing it on the head of the molecule.

- *vir D* is responsible for the synthesis of an endonuclease acting at the borders and involved in the liberation of T-DNA. Sometimes two genes are described, *vir D1* and *vir D2*.
- *vir E* codes for the synthesis of proteins that protect the T-DNA, gradually wrapping up the molecule during its transfer up to the nucleus of the plant cell and shielding it from probable attack by nucleases.
- *vir F* codes for the proteins that make them vastly superior in transferring DNA.
- *vir G* has also been described as responsible for the synthesis of a factor that activates the virulence.
- *vir H* has been described as responsible for enhanced tumorigenicity.

The expression of genes of virulence is generally triggered by molecules liberated by cells and tissues during an accidental injury (frost, trauma, waterlogging) or deliberate injury (scarification, excision of leaf parenchyma). The polyphenols often released by injuries are good candidates for representing intermediaries. It has, moreover, been demonstrated that the introduction of a supplementary copy of one of the *vir* genes in an agrobacterium activates the transformation.

A third region of the Ti plasmid, which like the *vir* region is never transferred, is occupied by genes for opine and octopine synthesis as well as by genes intervening in the conjugation between bacteria, a property that is exploited in genetic engineering.

There are also genes located on the bacterial chromosome, not on the plasmid, that intervene in the processes of bacterium-plant recognition and thus participate in the fixation of the agrobacterium to the plant cell.

As soon as the Ti plasmid is transferred, it replicates the missing strand of DNA and the remaining DNA strand serves as a template.

2.3.3. Controlled genetic engineering

There are two major modalities of transformation of plant cells. *Indirect transformation* involves the biological vectors, agrobacteria, which the experimenter uses to transfer the gene. *Direct transformation*, without an intermediary, often uses rather violent physical techniques to introduce the DNA carrying genes to be transferred into the plant cells (Fig. 2.15).

a) Indirect transformation

➤ Indirect transformation via *Agrobacterium*

Crown gall disease has inspired molecular biologists to develop an

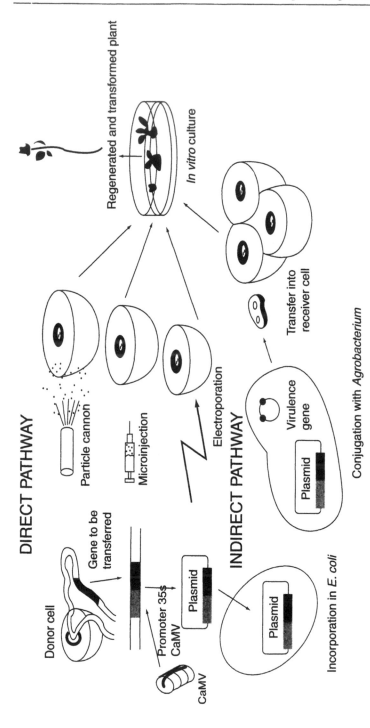

Fig. 2.15. General outline of the two possible pathways of gene transfer for modifying the genetic architecture of a plant cell. Direct transfer is based on three fundamental techniques: particle cannon, microinjection, and transfer by electroporation. Indirect transfer is based on *E. coli* and especially agrobacteria.

indirect method of transformation of plant cells and consequently the regeneration of plants.

The co-integration method

The first technique used was that of *co-integration* (Fig. 2.16). It had to meet two requirements: to reduce the size of plasmids to be handled in relation to the initial Ti plasmid and to suppress oncogenes in order to annul the disastrous effects of this plasmid on the development of tumour in the plant.

To meet the first requirement, the solution was to integrate the T-DNA into a standard cloning vector in *E. coli* and then to insert nearby, in the same vector, the DNA fragment to be transferred. The vector is then introduced in *Agrobacterium* containing its own Ti plasmids. With a little luck a homologous recombination will occur in *Agrobacterium*, resulting in the transfer of the construction realized into a wild Ti. This reprogrammed *Agrobacterium* in turn transfers its recombinant Ti to the plant.

To meet the second requirement, the *onc* part of the Ti plasmid must be deleted. When the deletion is successful, the result is a *disarmed Ti plasmid*.

The binary vector method

The technique that has just been described is now rarely used. The preferred technique is the *binary vector* method, which corresponds to the coexistence of two types of vectors in the cell (Fig. 2.17). One is a disarmed Ti plasmid still possessing its virulence genes and the other is a vector carrying DNA to be transferred inserted between the two typical borders of a Ti plasmid. This arrangement was conceived when it was demonstrated that the transfer results from independent, although complementary, activities of the T-DNA and the rest of the Ti plasmid. The efficiency of such a system has been verified chiefly in a large number of Dicotyledons. The quantity of DNA transferable by this technique is limited but sufficient to contain several genes in a series. On the other hand, it has not been possible to transfer large DNA fragments such as YAC or fragments derived from pulsed field gel electrophoresis.

The excision of oncogenes leaves a space large enough to be occupied by exogenous DNA representing genes to be transferred. In most cases there are two genes, one being the *target gene* and the other, placed in tandem with the first, being a *marker gene* or *reporter gene*. This second gene serves as a label for the first because it is easier to identify by one of the properties of its expression: gene for resistance to an antibiotic, gene coding for the synthesis of an enzyme intervening in a reaction the final

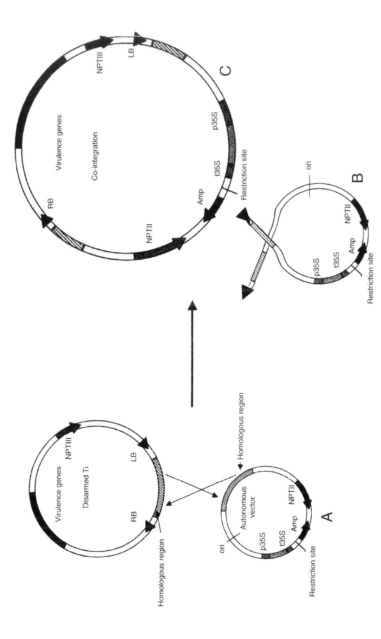

Fig. 2.16. Summary of co-integration technique. By means of the presence of homologous regions, the disarmed Ti plasmid and the vector carrying the gene to be transferred (A) fuse at the level of the T-DNA to form a large vector plasmid (C). In (B), a detail of the mechanism of co-integration of intermediate vector inside the T-DNA by a double section at the level of homologous regions, followed by ligation.

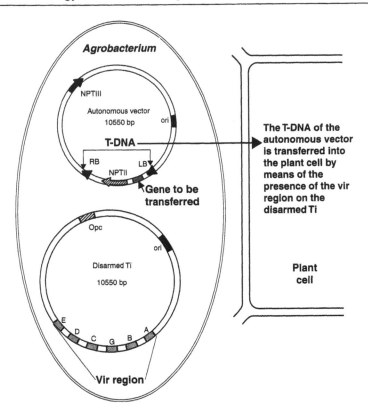

Fig. 2.17. Binary vector technique. The disarmed Ti vector and the autonomous vector, carrying right and left borders as well as the gene to be transferred, remain independent. The disarmed Ti allows the bacterium to express its virulence against the plant cell (specific recognition) and the autonomous vector transfers its T-DNA.

product of which is coloured (GUS gene or gene for ß-glucuronidase from *E. coli* and giving a blue reaction) or leading to an emission of fluorescence (e.g., luciferase in firefly). Resistance to an antibiotic is most frequently used because it allows for a highly effective system of selection. The resistance gene chosen is quite often resistance to kanamycin or neomycin phosphotransferase (*nptIII*), the gene for which comes from a bacterial transposon (bacterial resistance) or by insertion of the coding region of the gene of aminoglycoside transferase of type II (*nptII*), also from a bacterial transposon but conferring a resistance on plant cells. These resistance genes are inserted between the regulatory sequences of the gene of nopaline synthase. Resistance to other antibiotics is also used (ampicillin, hygromycin) as well as resistance to herbicides used in agriculture such as *basta* or *glyphosate*. These marker genes are placed between the border sequences and are thus transferred into the plant cells. They are highly useful in selecting cells that are really transformed out of all the cells subjected to a transformation operation, while the reporter genes serve most often to follow the activity of a promoter in terms of location as well as its level of expression. Among the plasmids most often used are Bin 19 and Bin plus (Fig. 2.18). Some researchers presently favour a new method of selection of transformants called MAT (multi-auto-transformation vector system) that uses oncogenes of *Agrobacterium* (*ipt* or *rol* genes). These oncogenes are inserted in specific recombination sites, so they can be withdrawn after selection of transformants and thus help us to obtain GMOs that are difficult to detect.

The *target gene* is the one in which the transfer must confer the desired new property on the plant. This could be a property the plant does not generally possess (bacterial, viral, or animal gene) or a property the plant possesses but at a level considered insufficient and not suitable. In the first case, the property acquired will be relatively easy to detect. If we take the example of a gene for luciferase in the firefly inserted in a plant, it can be identified by adding luciferin and ATP to the culture medium. An oxidation reaction ensured by the enzyme and consuming energy, i.e., concomitant production of AMP, will be accompanied by an emission of photons. Contact with a surface or a photosensitive film allows highly precise location of the sites of photon emission and, consequently, the presence of transformed cells. Note that the location of a gene does not alter in any way the integrity of cells, as sometimes happens with other reporter genes.

If the gene is already present but its level of expression is too low, it is possible to *overexpress* it by placing the coding part under a *strong*

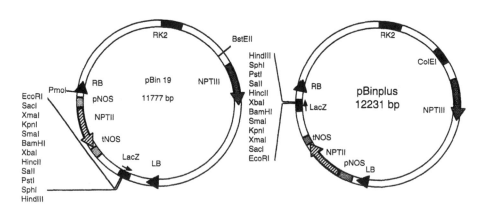

Fig. 2.18. Simplified map of Bin 19 and Bin Plus plasmids with their marker genes in bacteria (NPTIII) and in the plant cell (NPTII).

promoter, i.e., a promoter that is not sensitive to control signals from the plant genome. An example is the promoter of a gene coding for RNA of cauliflower mosaic virus or CaMV, which is capable of ensuring the transcription, at a high level, of a gene placed downstream if that gene is inserted in phase, i.e., at a suitable distance, expressed in number of base pairs. This promoter, called CaMV 35S, is the most frequently used but there are others, such as the promoter of the gene for nopaline synthetase, ubiquitin of maize, or those that come from genes of heat shock proteins (Fig. 2.19).

On the other hand, we may want to reduce or even eliminate the transcriptional activity of a gene. Besides deletion, which is not always easy to accomplish, biotechnologists have often resorted to insertion of an *antisense gene*. The coding part of the gene that needs to be exploited to a lesser degree is placed in a *reverse position* under the control of a strong promoter, which leads to the formation of messenger RNAs that are found to be complementary to messengers formed by the correctly oriented gene. It is generally acknowledged that the two mRNAs, oriented in the two complementary directions, can pair and thus form dimeric molecules that cannot be exploited by the translation system of the cell. No protein corresponding to these genes can be produced. A low

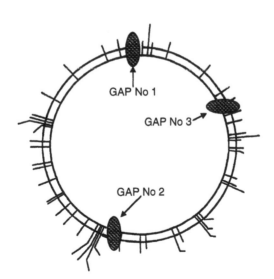

Fig. 2.19. Map of genome of the cauliflower mosaic virus (CaMV) indicating its three characteristic "gaps" and numerous restriction sites. The CaMV promoter 35S, used in transgenesis, is the promoter of the longest transcript of the virus. The specific search for this promoter in the seed cells is one of the techniques presently used for detecting contamination of seeds by GMOs.

level of expression can be explained by an incomplete pairing of messengers if the class of antisense is produced in a smaller quantity than the other (Fig. 2.20).

Gene transfers can only be considered successful if there are a certain number of controls. Sometimes, despite a well-planned manipulation, there is no control for correct interpretation of the results recorded. Resistance to an antibiotic is a serious primary indication of the presence of a gene but it must be complemented by controls of the presence of a DNA fragment containing the gene, using, for example, primers upstream and downstream and separating on the gel the presence of DNA fragments of the desired molecular mass. A Southern blot and electrophoretic migration of DNA can be used in parallel with that of the control of known molecular masses. The Southern blot is followed by the search for messenger RNA to be able to confirm the activity of transcription of the gene (northern blot). The last operation consists in identifying the translated proteins by western blot or by

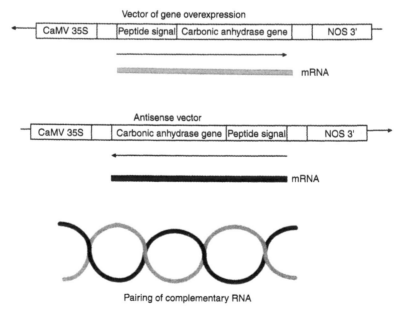

Vector of gene overexpression

| CaMV 35S | Peptide signal | Carbonic anhydrase gene | NOS 3′ |

mRNA

Antisense vector

| CaMV 35S | Carbonic anhydrase gene | Peptide signal | NOS 3′ |

mRNA

Pairing of complementary RNA

Fig. 2.20. Antisense strategy. The coding sequence of a gene is inserted in the reverse direction in relation with the strong promoter meant to control the expression of it. This gene codes for the synthesis of messenger RNAs, which are complementary to the messenger RNA of the gene in normal position. It is conceivable that the two categories of messengers could hybridize, thus preventing the translation enzymes from exploiting this information.

confirmation of a foreseen catalytic activity if it is an enzymatic protein, as is often the case. Some researchers then isolate the gene by restriction enzymes and again sequence the entire fragment. Such controls also allow us to know the immediate environment the gene benefits from during its insertion and to verify at the time the possible existence of several integrated copies of the gene at the same site. Among the other indispensable controls is the presence of identical plants, transformed with the vector in which the target gene is missing but the gene for resistance to the antibiotic is present. The differences recorded will be more certainly those introduced by the gene than those that result from transformation techniques. Other controls are identical plants having undergone all the operations of transformation for which the transformation medium was rigorously free of agrobacteria (distilled water controls).

The number of target genes introduced in vectors and then transferred into plants is very large and the list increases every day. An examination of the list of transgenic plants for which assays have been authorized by the Biomolecular Engineering Commission shows that there has been almost an exponential growth for several years. Most of these genes belong to the category of *genes of resistance* to products used in agriculture (herbicides, antifungal compounds, insecticides, antibiotics). A second, smaller category of target genes comprises metabolism genes, such as *metabolism of lignins* (antisense of OMT or O-methyl-transferase gene) to obtain less woody forage plants that are more easily digested by domestic animals or yield more fibre for the paper industry. Some genes are involved in the *metabolism of fatty acids* and can be manipulated to obtain plants that synthesize a large quantity of a particular fatty acid: with a short carbon chain for the energy sector, with a medium chain more or less saturated for the food sector, or with a long chain for plastic-coating industries. The *metabolism of nitrogen and its control* in plants constitute one of the more popular pathways of genetic engineering. In Chapter 3 we will look at some genes for *metabolism of isoprenoids*. Some genes intervening in the *respiratory chain* (e.g., genes for cytochrome synthesis) or in *photosynthesis* have been not only isolated and characterized but also transferred. We must also mention the *animal or human genes* that have been transferred into plants to produce a molecule with therapeutic uses (albumin, haemoglobin, immunoglobulin, human GAD proteins, antibodies of antiadhesine of mouse, hirudine of leech). A list of examples, by no means exhaustive, is given in Fig. 2.21.

In this way a new agriculture has developed, entirely directed towards the production of molecules meant for the bioindustries of health: *molecular farming* or *molecular pharming*, still experimental in

Sub-unit β of the heat-labile enterotoxin of *E. coli* Gastric lipase of dog Various monoclonal antibodies Encephalin Factors of epidermal growth Interferon β	Human seric albumin Human haemoglobin Rabies glycoprotein Plantibodies (surface markers of cancerous cell) Leech hirudine Erythropoietine (EPO) Cholera toxin

Fig. 2.21. Some molecules used therapeutically, products of transgenesis in plants. In 2001, some 34 proteins of pharmaceutical interest were listed, all products of genetic engineering.

some American farms but undoubtedly developing rapidly in other western countries. These structures depend, however, on only about 20 companies worldwide.

A typical procedure of genetic transformation of a plant

Practically speaking, a typical procedure of transformation is based on the following steps:
- 1. The operator must have available techniques and material indispensable to axenic cultures (autoclave, pasteurizer, systems of filter sterilization, 0.25 to 0.45 μm filters, glass containers, refrigerated centrifuges, etc.).
- 2. The operator constructs the transfer vector by combining several plasmids: one carries the gene to be transferred and another, such as the LBR19 plasmid, has unique restriction sites between the promoter CaMV 35S and its terminator plus resistance to an antibiotic (ampicillin). A third, for example Bin19, is capable of replicating in *E. coli* as well as in *Agrobacterium*. It also has resistance to kanamycin under control of regulatory sequences of the gene for nopaline synthetase, the right and left borders of T-DNA, and a polylinker that offers different possibilities of insertion of the target gene. A combination of these three plasmids using restriction enzymes and ligation allows us to construct the transformation vector *in vitro*.
- 3. The experimenter has two colonies of bacteria: *E. coli* cultured on an LB (Luria Bertani) medium at 37°C and *Agrobacterium* cultured on an AP (Annick Petit) medium at 28°C.
- 4. The *E. coli*, made competent by heat shock, is transformed by the plasmid DNA containing the transformation vector. The bacterial colonies transformed are cloned as a function of acquired resistance to antibiotics. Separation on gel electrophoresis allows us to subclone the colonies containing the recombinant plasmid.

• 5. The transformation vector must then be transferred from the *E. coli* to the agrobacteria. The transfer vector generally cannot be directly transferred. In this case, a *triparental conjugation* is used, which puts together, in addition to the donor and receiver colonies, a third partner represented by a new colony of *E. coli* possessing a particular plasmid known as *mobilizer*, which has functions of transfer. This colony is called a *helper*.

• 6. The experimenter must have plant material and know the techniques for *in vitro* culture of this material, especially culture mediums of the MS type. There will be the basic or control medium and the same medium with cefotaxim, an antibiotic designed to limit the proliferation of agrobacteria once they have fulfilled their transfer function. A third medium contains the single antibiotic kanamycin, designed to eliminate all the cells that have not received the transfer vector. The plant material to be transformed is generally made up of leaves or blade segments from young leaves sterilized in a sodium hypochlorite solution followed by rinses with sterile water (and sometimes a brief alcohol rinse).

• 7. From the sterile material, the experimenter then cuts discs or circles from the leaf blades (e.g., tobacco) and then co-cultures them with the agrobacteria for periods of about 10 minutes. After being rinsed, the leaf discs are cultured on a medium with cefotaxim for 2 to 3 days in the dark. During this period, the processes of transformation occur chiefly in cells close to the periphery of the disc and near the cells destroyed or altered by the sectioning. Also, during this period, the processes of cell multiplication essential to regeneration must be initiated.

• 8. The leaf discs are then transferred to a medium containing kanamycin in order to eliminate all the cells that have not been transformed, which are always the most numerous. Several weeks may pass before small islands of chlorophyllian cells are distinguished in some points at the periphery of the discs. The other cells lose their chlorophyll and die. The discs that manifest good indexes of regeneration are periodically subcultured on new mediums. It is often necessary to transfer these young regenerants to mediums without hormone so that a root apparatus appears and develops.

• 9. Plants regenerated on kanamycin are supposed to possess the gene to be transferred. However, that can be verified only after numerous controls have been operated with respect to the presence, activity, and integration of the character acquired, which controls have been described elsewhere.

• 10. The plants thus verified are taken out of the tubes to the greenhouse after careful acclimatization, as is done with all transgenic plants. They are cultured up to flowering so as to carry out the last control of heritability of the character.

➤ Other indirect transformations

We know that indirect transformation by *Agrobacterium* does not work in all plants and particularly in the large majority of Monocotyledons. It is possible in such cases to use viruses, as do biotechnologists working on the transformation of animal cells. These viruses are not as easy to manipulate, even though they have some advantages, in that they invade a much larger number of cells and sometimes even act as systemic invaders. Many plant viruses have an extended range of hosts and, consequently, the same virus can be used for a large number of plant species. Some of these viruses have been manipulated to become vectors of gene transfer in plants, for example, tobacco mosaic virus (TMV). These viral vectors are present in the form of plasmids containing the complete sequence of the cDNA, a marker gene that helps to achieve their cloning in a bacterium, and the target gene placed upstream of the promoter of the viral capsid protein. Geminiviruses, viruses with DNA in which the two chains do not pair, can also be used. One has an infectious character and the other can replicate in plant cells. The simultaneous presence of the two molecules is, however, indispensable to their activity. Part of the DNA molecule coding for proteins of the envelope can be deleted and replaced by the transgene. This part should be small in order to result in transformation of the host cell. This technique has given few results and has been in some cases advantageously replaced by direct methods of transfer.

b) Mutations by insertion of T-DNA

The insertion of T-DNA bearing a transgene may be accompanied by the mutation of a character that has no relation with the expression of the transgene. For example, *Arabidopsis* transformed by a T-DNA carrying the gene that confers on the plant resistance to an antibiotic (kanamycin) showed an unexpected phenotype with respect to size (dwarf plant) or development of floral organs. It was supposed that the point at which the T-DNA was inserted caused a disruption of the gene, leading to the appearance of a corresponding mutant phenotype. Verification of this hypothesis is difficult and requires a study on crosses during which the wild type can re-establish itself by complementation. This technique of random mutagenesis can be particularly valuable in locating genes or complex genetic functions such as the development or morphogenesis of organs.

c) Direct methods of genetic transformation

In agronomy, the importance of Gramineae and other Monocotyledons and the difficulty of transforming them by indirect methods have motivated the development of *direct methods of transfer*, i.e., methods that do not used a living organism as intermediary. The methods proposed are not all equally valid or efficient but they can complement each other to the point that they now make valuable contributions to indirect techniques in all plant biotechnology laboratories.

We have already cited the technique of protoplast fusion with liposomes by means of polyethylene glycol solution in the culture medium. These liposomes could contain DNA fragments that are large enough to contain one or several genes. The probability of transfer of this DNA into the nucleus of the plant cell is, however, extremely low on account of the presence of numerous nucleases and the absence of nuclear mailing of the transgene in the cytoplasm of the host cell. To this should be added the very great difficulty of regenerating plants from protoplasts, particularly cultivated plants. This technique has not therefore been developed further. Three other techniques have given much more favourable results and we will look at them in turn:

- microinjection,
- electroporation or electrotransfer
- biolistics and agrolistics.

We will subsequently discuss some variants or combinations of these techniques with each other as well as with indirect methods of transformation.

➤ Microinjection

Microinjection is not a highly original technique in that it has borrowed most of its experimental protocol from research in electrophysiology. For all that, it is not an easy technique to implement. The experimenter must have a micromanipulator associated with a inverse light microscope, all placed under a laminar flow hood because it is essential to maintain rigorously aseptic conditions throughout the manipulation (Fig. 2.22). Microinjection must be conducted on protoplasts in culture and individually for each protoplast to be transformed. Generally the yield is very low, especially when considered in relation to the considerable attention required and investment in technical work.

In practice, a protoplast is isolated from a culture, placed in a sterile microchamber, and held at the end of a glass cannula with a diameter much smaller than that of the protoplast (Fig. 2.23). It is held by means of a slight depression on the cannula by the micrometric withdrawal of the piston of a syringe connected to the cannula by a small catheter. Fixation is considered particularly favourable when the nucleus of the protoplast is found opposite to the point of fixation. In effect, the

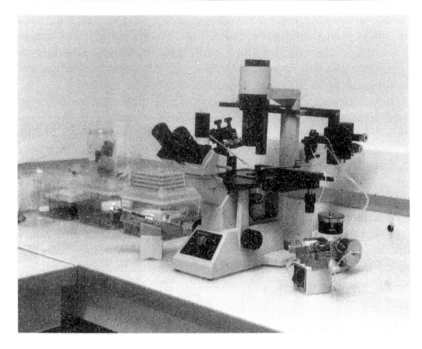

Fig. 2.22. Microinjection equipment using micromanipulator, in which the pneumatic controls can be seen at right. Microinjection is carried out under visual control through an inverse microscope.

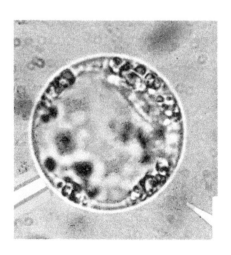 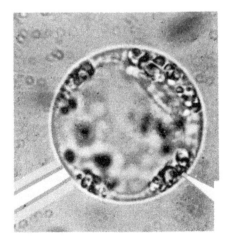

Fig. 2.23. Two views of the microinjection of a protoplast using a microneedle (right), while the protoplast is provisionally immobilized by a cannula.

protoplast, in the region near the nucleus, must be brought into contact with the tip of a glass microneedle, by means of a micrometric system of displacement. The microneedle is in fact a microelectrode containing the DNA that is to be injected into the protoplast. It is connected by a catheter to a second syringe on which the micrometric advance of the piston imposes a slight pressure intended to release the contents of the microneedle into the nucleus of the protoplast. The protoplast is then released, cultured, and followed up to the regeneration of the plant. This particularly delicate technique was developed by an American scientist, Croshaw, who microinjected about 100 protoplasts per day with a success rate of more than 60%. However, very few laboratories have so far succeeded in adopting the technique routinely and it still essentially belongs in the experimental domain. For example, no transgenic plant obtained by this technique has yet been registered with the Biomolecular Engineering Commission.

➤ Electroporation or electrotransfer

Electroporation is also not an original technique in itself since it is presently used largely in laboratories working on microorganisms, bacteria and yeasts. It is only an adaptation of essential parameters, duration and intensity of electric shock, to the characteristics of eukaryotic cells, particularly of protoplasts of plants. The protoplasts can be transformed by adding a concentrated solution of plasmidic DNA containing the insert to the culture in a chamber with electrodes between which an electric field is created (200 to nearly 1000 V per cm, for a few microseconds to milliseconds). The manufacturers of electroporators foresaw these extensions but the dissemination of the equipment has been hampered by the high price of installation and the low performance of more affordable equipment. Home-made and do-it-yourself initiatives have been developed in some laboratories and have given remarkable results. With respect to research on plants, the laboratory of Professor Georges Ducreaux at the University of Orsay in France deserves special mention. It has innovated profitably in this domain and its studies serve as a foundation for many other do-it-yourself products (Fig. 2.24). One of the qualities of the mechanism comes from the type of electrode recommended, which makes it possible to follow the different phases of the operation under the microscope. Many genetic transformations have succeeded in plants inaccessible to the indirect *Agrobacterium* system. Even though the yield is relatively low, the considerable number of protoplasts that have been treated by the operation renders the technique quite useful. This technique was used to obtain transgenic maize and rice.

Fig. 2.24. Equipment for electroporation of plant cells. At right is a generator of high frequency current used to align the protoplasts and surmounted by the electroporator itself. The fusion operation is observed under inverse microscope.

➤ Biolistics

Biolistics is an original technique that has been developed particularly in the plant biotechnology sector. It is a method of "bombarding" plant cells or tissues using metal microbullets coated with DNA and projected by a "gun". This gun in fact predates the technique, i.e., it dates from about 1986. It consisted of a 22 LR calibre gun and cartridges charged with powder and tungsten microbullets of about one micrometer in diameter. The target was a leaf spread out on filter paper. The shot transmitted kinetic energy to the microbullets, which then reached a speed of around 400 to 500 m/sec, and the cells hit by the microbullets died. However, at the periphery of the impact, cells belonging to the epidermis or the underlying palissadic parenchyma received some projectiles without being killed. Among these surviving cells some received projectiles in the chloroplasts, others in the nucleus, and it was shown, especially using the GUS gene and genes coding for the synthesis of anthocyanins, that foreign genes could thus penetrate and express themselves. The technique evolved and the gun was replaced by a cylinder of gas under pressure, for example liquid helium (Fig. 2.25). This gas compressed at about 200 atm is brought to a pressure of 5 to 8 bar by a regulator that allows a small airlock, a small chamber closed at its lower end by an electric valve, which can regulate

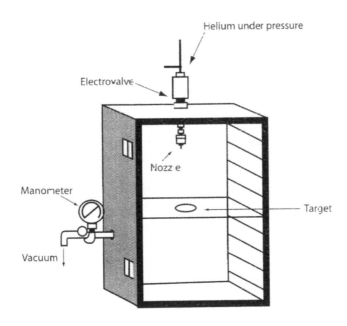

Fig. 2.25. Equipment for genetic transformation by a particules gun with tungsten microbullets. The plant matter is shot under a partial vacuum of about 25 mm Hg. At the moment of the shot, an electrovalve on top of the container releases helium compressed at 7 bar contained in a small airlock. The gas itself comes from a cylinder of liquid helium that has a regulator connected to the airlock.

the duration of opening. The valve directly leads to an analogous equipment with a filter-port having a metal grill on which the DNA-coated tungsten microbullets to be transferred are placed. The helium, thus suddenly released, projects the microbullets towards the target, a petri dish containing the plant material (e.g., leaves, leaf segments, leaf discs, cells in suspension, seeds). The equipment carrying microbullets and the target are placed in an airtight glass container, in which a partial vacuum is created, which facilitates the trajectory of bullets and effectively maintains conditions of sterility. The whole apparatus is placed under a laminar flow hood. The parameters that the experimenter can easily vary are the outgoing pressure of the regulator, the quantity of projectiles, the concentration of DNA, and the distance from the valve to the target.

The results obtained with this technique belong to three categories:

- The DNA carrying the transgene penetrates the cell but is not integrated in the nuclear genome. This is the most frequent case and only a transient expression is found. Such a result could be particularly useful for verifying the functioning of a gene, knowing its products of transcription and translation, but it cannot be extended and the cell is not genetically transformed.

- The DNA carrying the transgene penetrates one or several chloroplasts of the receiver cell. There may be genetic transformation of the organelle. This has been demonstrated in the case of transfer of a gene coding for synthesis of 16S RNA in tobacco but carrying a mutation conferring resistance to an antibiotic (streptomycin). Cells with genetically modified chloroplasts have been successfully regenerated with some plants carrying the transgene and resistance to the antibiotic. However, such results are still rare and the yield is probably very low. Nevertheless, it is the only technique that can realize the transformation of such organelles and it is a useful and original approach that will probably be further improved.

- The DNA carrying the transgene penetrates the nucleus of the receiver gene. It may find in the genome a region partly complementary with the sequence of the transgene and make an exchange of the *crossing-over* type. The cell is thus genetically transformed and the character could be transmitted to its descendants. This case is not frequent but some successes have been reported, especially in maize, with genes coding for the synthesis of storage proteins in seeds or with genes conferring resistance to a herbicide used in agriculture. The technique is used for plants that are difficult to transform using the *Agrobacterium* system.

It should be noted that the cell receiving the projectile(s) is not itself directly the source of regeneration because it is often killed by the injuries suffered during the shooting. It needs to divide before it dies and the sister cell, which inherits a copy of the transgene but has not suffered the impact of microbullets, is the origin of the regeneration process.

Biolistics is developing and continues to evolve. Some users have recently associated it with indirect transformation by agrobacteria. The cannon thus serves essentially to cause microinjuries in the tissues. These injuries, caused by microbullets or small metal shavings, allow agrobacteria to invade the tissue more easily and vigorously and thus increase the productivity of the transformation. This is called *agrolistics*. This technique can also be used for genetic transformation of organelles of the cell, plastids and mitochondria.

Plant biotechnologists are today faced with a varied and effective range of techniques of genetic transformation, a range that will probably be enhanced in the years to come. The limiting factor, with respect to agronomic applications, most often comes from *in vitro* regeneration techniques, a vast domain in which progress, even though it is certain and promising, is nevertheless relatively slow. The difficulty of regeneration by the somatic route still represents a technological hurdle for some plants of agronomic interest, whether for market gardening or horticulture. Another problem arises from the presence of marker genes, often genes of resistance to chemical or biological products. They are indispensable for the selection of transformed tissue, but it is preferable to suppress their presence during the transfer of the transgenic plants into commercial production.

2.4. PLANT BIOTECHNOLOGY AND OTHER BIOTECHNOLOGIES

Plant biotechnology presents only one part of a range of biotechnologies available to researchers studying and seeking to resolve agronomic and bioindustrial problems.

Biotechnologies that involve microorganisms, particularly prokaryotes, are the most accessible and easiest to implement and, because they have been used longer, a greater amount of information and perspective is available. A deeper knowledge is necessary for manipulations of eukaryotes, since there is often a phase of genetic amplification or transfer that relies on these microorganisms. The transformation procedures are more homogeneous, depending largely on electroporation and more rarely on metallic ions.

Animal biotechnologies follow a similar progress but the transformation vectors are not the same. There is no bacterial model equivalent to the *Agrobacterium* system and, moreover, animals do not have the same regeneration capacity as plants. For indirect transformation, the vectors used are most often viruses but the viruses must be non-pathogenic and lead to a stable integration of the transgene in the host cell: two conditions that considerably reduce the choice of possible candidates. It is also possible to use retroviruses, which have the merit of correctly transporting the transgene until its integration in the genome. In indirect transgenesis, microinjection is the most popular and convenient technique in most cases. However, not all the cells of an individual can be injected. For effective transformation, the researcher must use the egg cell or any embryonic cell that still has all its potentiality. Totipotentiality is known to be very short-lived in most animals, limited to the very early stages of embryo development. Despite all the advances, the transformation success rate is still quite low, about 1 out of 10 to 1 out of 100 embryos manipulated in the higher animals. It is high in animals with external fertilization, such as fish, batrachians, and echinoderms. There have also been attempts to transport DNA by injection into a spermatozoid, which then serves as a vector for a DNA fragment without attempting to integrate itself in the genome of the gamete.

In mammals, the mouse is the standard animal for transgenesis. The attempts and successes are at present essentially directed towards fundamental research, particularly on the specific functioning of genes and their modalities of control. Nevertheless, there are already some applications: as a source of recombinant proteins secreted in the blood or in milk, for protection against some diseases (Mx genes of influenza, production of antibodies, interleukin), or for improvement of performance of some animal races. There are also interesting applications in pharmacology, such as the study of the role of oncogenes and processes of tumour formation.

In this context, biotechnology and the mastery of genetic engineering in plants do quite well. The techniques in this sector seem today to be the most advanced and carefully controlled, albeit with some disadvantages due to its pioneering character: it finds itself at the frontline in the face of criticism from opponents, it must answer many questions from the public, it must set up regulations and legislation adapted to GMOs, and it must participate in the major debates of our time on the utility or futility of this field.

FOR FURTHER READING

The journals: *Biofutur, Scientific American, La Recherche, Science et Avenir...*

Chawla, H.S. 2002. *Introduction to Plant Biotechnology*, 2nd ed. Science Publishers, Inc.

Chun, Y. Huang, Ayliffe, M.A., and Timmins, J.N. 2003. Direct measurement of the transfer rate of chloroplast DNA into the nucleus. *Nature*, 422, 72-76.

Dixon, R.A. 1987. *Plant Cell Culture*. IRL Press, Oxford.

Fowke, L.C., and Constabel, F. 1984. *Plant Protoplasts*. CRC Press Inc., Boca Raton.

Margrara, J. 1982. *Les Bases de la Multiplication Végétale*. INRA.

Old, R.W., and Primrose, S.B. 1985. *Principles of Gene Manipulation*. Blackwell Scientific Publications, Oxford.

Pierik, R.C.M. 1987. *In vitro Culture of Higher Plants*. Martinus Nijhoff, Dordrecht.

Reinert, J., and Yeoman, M.M. 1982. *Plant Cell and Tissue Culture*. Springer Verlag, Berlin.

Scriban, R. 1996. *Biotechnology*, 6th ed. Lavoisier, Paris.

Thorpe, T.A. 1981. *Plant Tissue Culture: Methods and Applications in Agriculture*. Academic Press, Inc.

Vasil, I.K. 1984. *Cell Culture and Somatic Cell Genetics in Plants: Laboratory Procedures and Their Applications*. Academic Press, Inc.

Chapter 3

Applications of Biotechnology and Genetic Engineering

3.1. EXAMPLES OF THE IMPACT OF BIOTECHNOLOGY AND GENETIC ENGINEERING ON AGRONOMY

3.1.1. Problems of choice and evaluation

Plant biotechnologies, particularly genetic engineering, have undoubtedly had an impact on agriculture because of the hopes and uncertainties that they have raised in the public mind. The media has also played a major role in enhancing this impact, as evident in the positions taken by associations, pressure groups, and politicians. This mobilization has brought about active debates involving producers, ecologists, distributors, consumers, and gourmets, especially when the producer is simultaneously an ecologist, gourmet, and necessarily consumer. Scientists, especially plant biologists, can at least console themselves that their field no longer induces polite indifference or condescension about their work with "little flowers", as was often the case even in the recent past.

Although the impact on public opinion has been relatively great, it cannot be explained by the direct consequences of these technologies, which have till now been relatively minor in our daily lives. Our environment is still essentially the product of natural selection, which we merely observe and analyse, without participating closely in it. The same will undoubtedly hold true for some time in the future. We are in a transition phase and any assessment in this field can only be provisional.

There are several means of making such an assessment. We can consider the plant and its various organs or its major metabolic pathways or even the products we take from it for consumption and economic activities. This transition period is not favourable for a choice of one method over another. We will attempt only to make several sketches while waiting for the future to make clear the more suitable choice.

3.1.2. Plants affected by biotechnology

In principle, all plants are involved in biotechnology; in practice, it is essentially the Phanerogams, which have an agronomic interest or are of use in horticulture or gardening, that are involved, besides the small Brassicaceae *Arabidopsis*. One indication about species involved in biotechnology can be drawn from the analysis of data published by the Commission du Génie Biomoleculaire (CGB). This commission periodically publishes the list of species, types of projects, and dossiers on field trials. There are also sources of information involving transgenic plants that are being experimented upon. Figure 3.1 shows the evolution of plants of agronomic interest involved in transgenesis for the years 1995 to 2001. Until 1997, anti-GMO movements had not yet become very active. From 1998 onward, the imposition of moratoria (especially for rape) interrupted this evolution, which had a purely economic origin.

We can also assess the importance of studies on transgenic plants from statistics known for the North American continent, from the size of areas occupied by the objectives pursued (Fig. 3.2).

Nevertheless, it is advisable to read these tables with some commentary if we are to have a more realistic idea of the relative importance of species with a biotechnological value. *Arabidopsis*, a model plant accessible to transformation techniques, has an importance that has no relationship with its economic interest. The score of tobacco, another model plant according to the applications in 1996 (62 dossiers), is much higher than the agronomic importance of this species would indicate, even though it is a plant with considerable economic weight.

A. 1995	B. 1997	C. 1999	D. 2001
Plant and number	Plant and number	Plant and number	Plant and number
Maize, 21	Maize, 26	Maize, 33	Maize, 10
Rape, 19	Rape, 19	Beet, 13	Rape, 4
Beet, 9	Tomato, 11	Rape, 11	Beet, 2
Tobacco, 8	Soybean, 7	Potato, 4	Poplar, 2
Tomato, 2	Cotton, 7	Sunflower, 3	Sunflower, 1
Soybean, 1	Potato, 5	Tobacco, 2	Potato, 1
Grapevine, 1	Tobacco, 2	Soybean, 1	Wheat, 1
Melon, 1	Melon, 1	Grapevine, 1	Sunflower, 1
Others, 3	Others, 4	Others, 3	Others, 3

Fig. 3.1. Examples of plant species involved in genetic engineering classified as follows: in A, C, and D according to the number of dossiers processed in Western Europe by the CGB or, in B, according to the number of programmes declared in connection with the DIGIP Database. In 2001, the consequences of the 1998 moratorium on GMOs were observed in an overall decline on research in Europe on these plants.

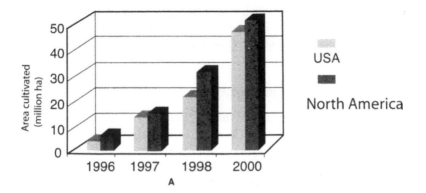

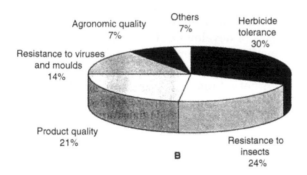

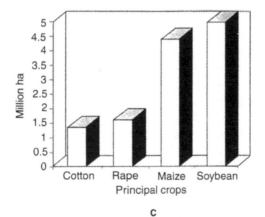

Fig. 3.2. A. Area covered by transgenic plants in the United States and on the entire North American continent (expressed in millions of hectares). B. Relative proportions of objectives pursued and, consequently, types of characters introduced. C. The four principal transgenic crops in the United States (for 1997).

For that same year, maize (with 102 dossiers) took first place, followed very closely by rape (101 dossiers). Rape poses major problems of dispersion of transgenes and for this reason was ranked lower in 1998 and 2001.

3.1.3. The whole plant and its organs

Does genetic transformation involve the whole plant rather than specifically some of its organs? There is still little to be said about this subject. Certainly, when a gene is transferred under a strong promoter, it is generally the whole plant that is involved. This choice is explained easily with respect to the acquisition of resistance or protection or when it is a question of intervention in an essential metabolic pathway, which is exerted at a comparable level in all parts of the plant. There are, however, a large number of studies at present that rely on spatial or sequential differential expression of trangenes, i.e., depending on the organs or different stages of development. It is thus possible to consider the structures that constitute anatomic or functional units: tissues and organs. In this respect, organs of great interest to researchers seem to be the leaf for its photosynthetic functions, the flower because of its role in reproduction, and the fruit and seed, for their economic aspects. Following these are the roots, stems, and secondary ramifications.

The *leaf* is primarily used because it is easy to remove without harming the plant. It is also often the organ that regenerates most effectively. Finally, it is in the leaf that the most important function of the plant is realized: the input of carbon and nitrogen into living matter. It is thus naturally chosen for study of the fundamental steps in these physiological processes.

The *flower*, as a whole, has mostly been the subject of studies about ontogenesis of its various organs. Genetic engineering has certainly contributed greatly to this sector and important discoveries were made at the end of the 20th century. It is probably one of the domains in which we can still expect the greatest progress. Apart from a considerable abundance of fundamental information, there have been breakthroughs in horticulture with the creation of flowers with multiple whorls and variations in flower colour. By this means petunias with red flowers were created by introducing the dihydroflavone-4-reductase gene from maize and other white flowers were created by blocking the transcriptional activity of the chalcone synthase gene using the antisense technique.

Control of *fruit* maturation is probably one of the very first applications of plant transgenesis to have reached the consumer's table. The maturation phase is known to correspond to changes in the nature of pigments, an increase in the level of soluble sugars, synthesis of specific aromas, and progressive hydrolysis of the pectic cement that connects the cells to one another. This hydrolysis is due to the activity of an enzyme called polygalacturonase. There is moreover a sudden production of ethylene, a gaseous hormone that plays an important role in communication between cells and synchronization of fruit maturation. There have thus been two courses for biotechnologists, who have been asked to slow this maturation in tomato in order to improve its ability to withstand transport and storage and to better regulate market movement. Obviously genetic engineering, particularly the use of antisense technology, was used and gave the desired results. The first variety of transgenic tomato with long shelf life was commercially produced in the United States in 1994 by Calgene and a second in the following year by the Zeneca company. In both cases, success was moderate more because of media campaigns brought on by opponents than by actual criticism of the organoleptic qualities of the fruit. These tomatoes do have their supporters and have qualities that are useful in the manufacture of concentrates, which are not legally considered transgenic products. Other fruits, such as melon, have also been genetically transformed for similar purposes. In banana and other fruits, assays have been carried out to introduce genes that trigger the production of vaccines, such as hepatitis B vaccine, by the fruit, with the ultimate aim of "vaccinating" whole populations in countries in which these maladies are widespread, especially developing countries. If such a product were to come to fruition, a vaccination campaign would be greeted with enthusiasm!

At the scale of *tissues* and *cells*, the primary meristems, the assimilator parenchyma, the vessels and their associated cells, the fibres impregnated with lignin, the trichomes or hairs, and the secretory organs have more particularly attracted attention. The filling tissues, lacunal parenchyma, collenchyma, and pith seem to be less important. Sometimes the organs are considered in terms of the physiological criterion of source-sink duality and special attention is given to the mediators of metabolite transport.

All the parts of an organ do not attract equal attention. In the stamen, for example, it is mostly the sporogenic tissue that is of interest, as well as tapetum cells, since there is a possibility of controlling sterility and fertility through this tissue. The stigmatal tissues of the pistil have also been considered important because of the role they play in the control of pollen compatibility. Most of the choices are clearly guided, even for researchers who fiercely defend the technology for its own sake, by economic considerations downstream. It is regrettable, for example, that biotechnological studies have practically ignored the pivotal group of Pteridophytes, which are very important for an understanding of the organization and functioning of higher plants. Algae have received slightly more attention, but it is true that they represent enormous food and industrial economic potential.

3.1.4. Major metabolic pathways

We do not intend here to analyse all the known metabolic pathways in plants and the impact that biotechnology has had on their understanding and control. We limit ourselves to some examples, keeping in mind that these pathways are all highly complicated and depend on one another. Pyruvate and acetyl coenzyme A, derived from glycolysis, constitute a hub from which the multiple pathways of biosynthesis radiate. Figure 3.3 gives a quick overview of these different pathways and their interrelationships.

a) The fatty acid pathway

All plants produce molecules of fat from glycerol in which the three alcohol functions are esterified by fatty acids. The so-called oleaginous plants, the best known of which are sunflower, rape, soybean, maize, peanut, and olive, produce large quantities of edible oils. Others, such as flax, *Camelina sativa* (L. Crantz) gold-of-pleasure, and ricin produce oils that are used more for industrial or pharmaceutical purposes. In the plant, the oils are generally stored in the seeds as an energy source for embryo development, but also to allow the plant to resist frost. The fatty acids are linear carbon chains with a length varying according to the number of carbon atoms (12 to 18 for edible oils, 20 to 24 for industrial oils, and longer for waxes) (Fig. 3.4). They also differ in the absence or presence of double bonds between two adjacent carbon atoms. In this case the number and position of double bonds within the molecule are an essential element of the interest in it.

The best-known fatty acids, because they are the most common, are palmitic acid (C 16) and stearic acid (C 18); they are called saturated because they lack double bonds. Variants of these fatty acids are obtained by introduction of double bonds under the action of enzymes called *desaturases*. For example, the δ-9-desaturase of chloroplast introduces a double bond between carbons 9 and 10 and transforms stearic acid into its variant, *oleic acid*, which is the major constituent of edible oils (Fig. 3.5). There is an entire series of desaturases whose specificity is linked to the carbon atom near which this enzyme introduces the double bond. Several genes coding for the synthesis of desaturases have been isolated and cloned (desaturases corresponding to double bonds in Δ-6, Δ12, Δ15) from mutants of *Arabidopsis* and, in some cases, by genic disruption following random insertions of T-DNA. The sequences of these genes have been established and they have been used to clone the genes having the same functions in oleaginous species

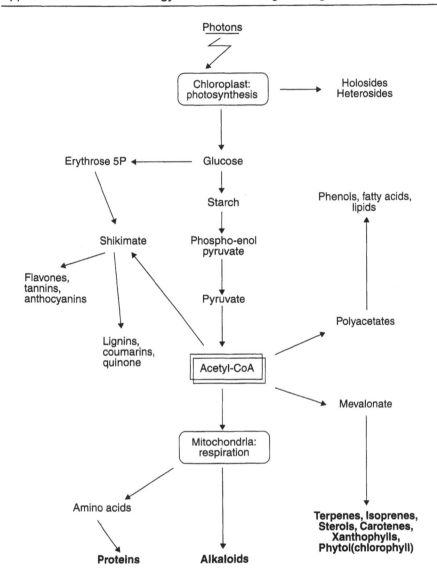

Fig. 3.3. Simplified diagram of the major metabolic pathways. The relationships between the metabolic pathways studied in the chapter are indicated by arrows.

(sunflower, rape, and soybean). In effect, it is observed that the sequences are relatively well conserved, from cyanobacteria to higher plants. However, some of these fatty acids are absent or very poorly represented in crop species, while they are indispensable to human metabolism. This is

	No. of carbons	Name	Symbol	Location
Short chain	4	Butyric	4:0	Butter
	6	Caproic	6:0	Butter
	8	Caprylic	8:0	Milk, palm tree
	10	Capric	10:0	Milk, palm tree
Medium chain	12	Lauric	12:0	Laurel
	14	Myristic	14:0	Nuts of nutmeg,
	16	Palmitic	16:0	palms, and many
	18	Stearic	18:0	plants
	18	Oleic	18:1Δ9	
	18	Linoleic	182Δ(9, 12)	
Long chain	20	Arachic	20:0	Peanut
	20	Arachidonic	22:4Δ(5,8,11,14)	Peanut
	22	Behenic	22:0	Various seeds
	22	Erucic	22:1Δ13	Rape
	24	Lignoceric	24:0	
	26	Cerotic	26:0	Waxes of plants,
	28	Montanic	28:0	bacteria and
	30	Melissic	30:0	insects
	32	Laceroic	32:0	

Fig. 3.4. Characteristics of major fatty acids classified according to the length of the carbon chain and their possible double bonds.

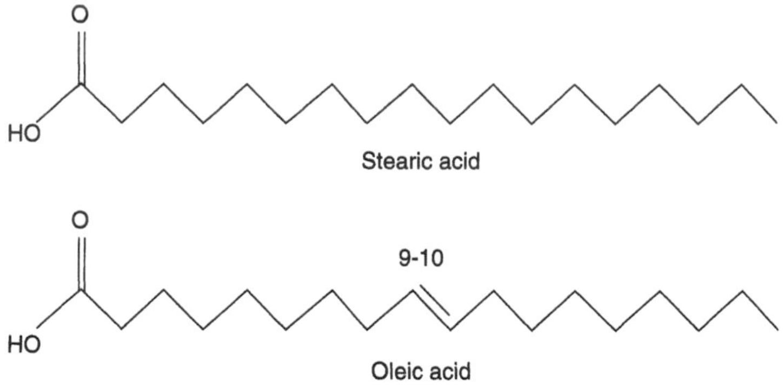

Stearic acid

Oleic acid

Fig. 3.5. Formulas of saturated stearic acid and its derivative, unsaturated oleic acid, with a double bond between carbons 9 and 10.

true of α-linolenic acid, precursor of the synthesis of human prostaglandins, absent from most edible oils but found in very high quantities in the lipid reserves of borage seeds. Unfortunately, this plant is not familiar to our cuisine and it could seem preferable to borrow just the gene and introduce it into sunflower or rape. The project seems entirely feasible since the gene has been transferred to tobacco.

The fatty acid industry has such economic significance that it directly influences agricultural activity by imposing its quality criteria and production standards. The various sectors of the industry have a wide range of needs with respect to the length of the carbon chain, the presence and position of double bonds, and the content of the initial material, i.e., most often the seed. For example, the sector producing oleine is adapted to an extraction of this product from seeds in which the oleine content ranges from 75 to 85%. Selection by conventional methods has allowed production of sunflower varieties that reach and sometimes exceed the required content. A similar level is much more difficult to reach in rape and there are several genetic engineering projects involving this plant.

b) The isoprene pathway

Isoprenes (or isoprenoids) represent a multitude of products with extremely varied functions and structures, most of which play a fundamental role in the physiology of all living things (hormones, membrane steroids, cell cycle). The biosynthesis pathway of these products comprises a certain number of steps from which are derived many other biosynthetic pathways, particularly in plants, for their primary as well as secondary metabolism (Fig. 3.6). Downstream, these products are important in agrofood and bio-industrial applications, in sectors as varied as agrochemistry (pesticides), dyes, perfumes, aromas, antifreezes, pharmaceuticals, and polymers (rubber).

The isoprene pathway is also peculiar in having in its various categories of living things a very large common trunk in its central part (Fig. 3.7) as well as some specificities in its upstream part (Prokaryotes on the one hand and Eukaryotcs on the other; Fig. 3.8) and in its terminal part (animals, fungi, plants; Fig. 3.9). These specificities are obviously widely used in all the discriminant applications, such as those of fungicides in which the treatments could have drastic effects on fungal sterols (ergosterol), while sparing plant sterols (phytosterols) or animal sterols (cholesterol).

In plants, the isoprene pathway leads to the synthesis of molecules indispensable to carbon metabolism (e.g., carotenes, xanthophylls, the phytol chain of chlorophyll), essential for growth and development (e.g., hormones of the cytokinin, gibberellins, and abscisic acid type), the constitution and functioning of membranes (phytosterols), respiration (quinines, cytochromes), and cell division (prenylate proteins). This same pathway leads to the formation of products that protect plants (terpenes, phytoalexins), aromas important for pollination (nectars), and numerous secretions (gums, essential oils, rubber). A large number of terpenes are used in the health sector (taxol, for example, an antitumoral extract of *Taxus* or yew). In addition, insects essentially use phytosterols to achieve the synthesis of ecdysone, their moulting hormone.

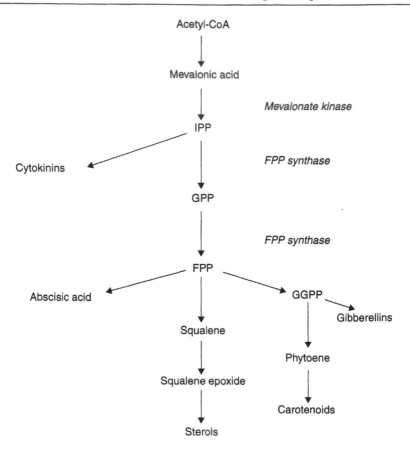

Fig. 3.6. The principal steps of the sterol biosynthesis pathway and derivative pathways of considerable importance in plant growth and physiology.

Mastery of the steps of the isoprene pathway is clearly of great interest to biotechnologists.

The most numerous and profitable biotechnological studies in this field have been conducted in fungi and yeasts. The metabolic steps are now well known and the large majority of enzymes intervening in this pathway are identified. Considerable advances have been made in the identification and characterization of genes that code for the synthesis of these enzymes, especially because of the availability of many mutants. Two approaches have been particularly well developed. The first is the use of inhibitors, often by competition, of terminal enzymatic steps that have given rise to an entire group of fungicidal products known as sterol

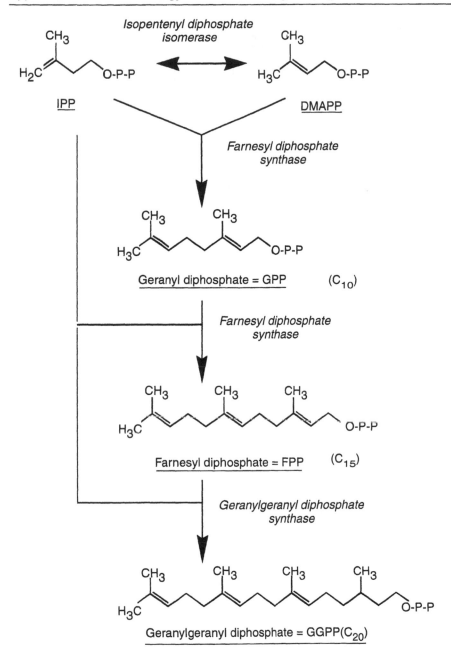

Fig. 3.7. Central part of the isoprene biosynthesis pathway, which is common to all living organisms and corresponds to the condensation of two or three molecules of isopentenyl diphosphate, which gives farnesyl diphosphate (FPP) and geranyl diphosphate (GPP) (S. Champenoy).

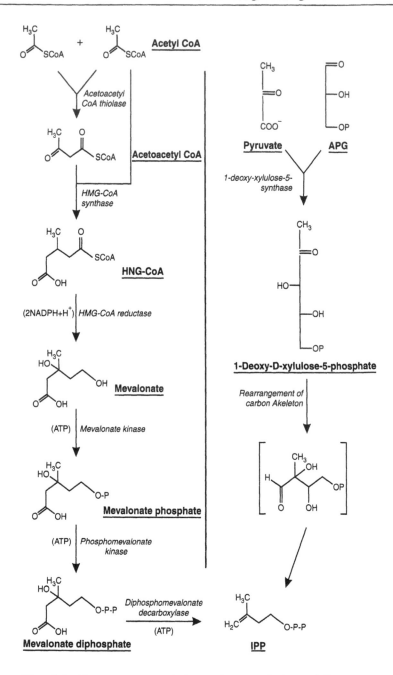

Fig. 3.8. Upstream part of the pathway, which is different in Prokaryotes (right) and Eukaryotes (left). The prokaryote route is also found in semi-autonomous organelles, plastids and mitochondria. The two pathways end in IPP (S. Champenoy).

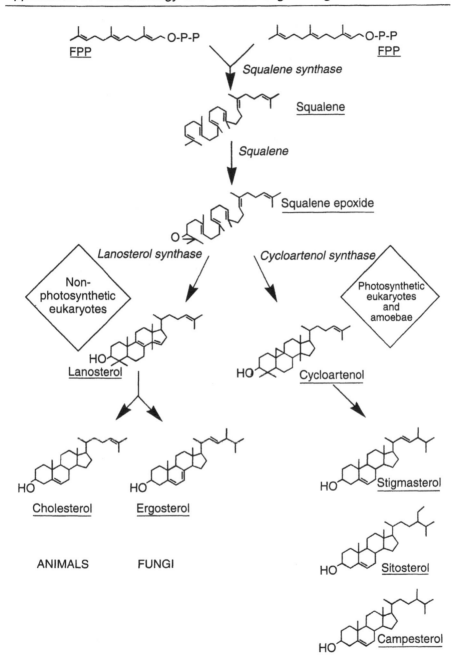

Fig. 3.9. Downstream part of the pathway, which is distinct in animals (cholesterol), fungi (ergosterol), and plants (phytosterols). The efficiency of antimycosic and antifungal compounds of the IBS type lies in these differences (F. Karst).

biosynthesis inhibitors (SBIs) used in agriculture as well as medicine (antimycosics). The second is the knowledge of yeast genes, which has allowed identification of a certain number of homologous genes in *Arabidopsis* (HMG CoA reductase, mevalonate kinase, farnesyl-diphosphate kinase, squalene synthase, phytoene synthase, etc.) and subsequently in other plants (tobacco, maize).

The cloning of these genes was followed by experiments to transfer them into plants of agronomic interest with a view to their overexpression or inhibition (antisense technology). Taking into account the implication of products of this pathway for growth, significant effects were recorded on regeneration capacity, growth rate, precocity of flowering, fruit maturation, and seed yield. These experiments in overexpression were valuable in determining the limiting steps and modalities of control normally exerted within the cell. For example, tobacco overexpressing farnesyl diphosphate synthase can produce quantities of carotene and phytosterols that are three to four times the normal level, thus conferring on the plant better protection against radiation and increased resistance to some specific inhibitors, according to S. Daudonnet. There are also transgenic plants overexpressing the mevalonate kinase gene, which thickens their stem, reinforces their root system, and accumulates impressive quantities of starch, according to S. Champenoy. Finally, plants overexpressing animal genes (e.g., HMG CoA reductase of hamster) have been successfully produced and have resulted in patents in studies about control of hypercholesterolemia (Amoco).

The interdependence of plants and the moulting of insects has not escaped the attention of researchers, who have seen a possibility here for controlling populations of predatory insects by controlling the phytosterol profile of cultivated plants. Projects in this field have already gone beyond the experimental stage. The possibility of regulating the quantity of hormone secreted by plants opens up significant agronomic perspectives.

c) The nitrogen supply of plants

Nitrogen levels are a fundamental subject since the input of nitrogen into the biosphere is indispensable for protein synthesis. Since the most abundant source is molecular nitrogen from the atmosphere, biochemists and microbiologists have long looked at the modalities of the transfer of atmospheric nitrogen into the biosphere, this step being ensured mostly by bacteria and cyanobacteria. Plants cannot directly utilize atmospheric nitrogen except when they make a symbiotic alliance with one of these prokaryotes. Leguminosae, particularly the family Fabaceae, represent the best example of symbiosis between plant roots and a nitrogen-fixing bacterium, *Rhizobium*. An analogous result is obtained by some trees, such as alder, with *Actinomyces*. There are many other cases and the drawing up of a complete inventory seems to be a distant prospect. Nitrogen fixation in bacteria is carried out by *nitrogenase*, the synthesis of which is coded by the nif gene. Experiments to introduce this nif gene directly into the genetic material of plants, particularly Gramineae, have often been proposed. Research programmes pursue these objectives but practical realization and commercial applications do not seem possible in the near future.

The large majority of plants extract nitrates from the soil by means of *nitrate reductase*, which, by the intermediary of the mitochondria-chloroplast pair, allows the plant to store nitrogen or to assimilate it in the synthesis of amino acids. Some transgenic tobacco varieties, expressing the gene coding for nitrate reductase synthesis at varying levels, have shown that it is possible to control the level of nitrates contained in the leaves. Other assays on overexpression of *nitrate reductase* under the CaMV 35S promoter have also been undertaken in tobacco in order to optimize the assimilation of nitrates. The increased utilization of nitrates, which normally accumulate in vacuoles, is accompanied by a reduction of their entry into cells. In parallel there is an increase in glutamine synthesis in the leaves. However, some difficulties have arisen and sometimes the results recorded were observed to be totally opposite to those expected. These events have been related to phenomena described as *cosuppression*. Some assays have been carried out on spinach. We can also hope that these experiments result in better management of nitrogen input through optimal use of fertilizers in agriculture, which would lead to better preservation of groundwater resources. However, it presently seems easier to introduce bacteria of the *Rhizobium* type directly into plant cells to obtain artificial symbiosis. Some success has been recorded in rice and rape on cultures of cells treated with 2,4-D and PEG in which the walls were partly digested. However, generally less than 10% of plants treated engage in this process of symbiosis. Besides, the symbiosis usually remains extracellular and the nitrogen-fixing activity in the cells is very low.

d) Other metabolic pathways

The metabolic pathways that have been at least partly controlled by biotechnology and genetic engineering are numerous and can never be reviewed exhaustively, especially in view of the rapid rate of development of these technologies. We shall mention the *carbon nutrition* pathway, i.e., photosynthesis itself, and the transport and distribution of assimilates. Rubisco, a key enzyme of this pathway, has often been the subject of these techniques and we now understand the mechanisms of synthesis and the movement of small subunits at the level of cytosol and chaperone proteins that regulate the union of different subunits. A certain number of genetic manipulations have also been undertaken with the aim of reducing, or even suppressing, its oxygenase function, which is said to be energy-intensive. Researchers have believed that the biosynthetic capacity of this enzyme can be increased by favouring its decarboxylase role. The major difficulty lies in the fact that its catalytic functions are carried out by large subunits and that these are synthesized in the chloroplast and consequently in a situation less favourable to transformation by means other than biolistics. The few successes obtained have had the merit of showing that this oxygenase function is indispensable, apart from the synthesis of amino acids, to the survival of the plant in particularly delicate situations of stress. The study of assimilate transport, the identification of transporters and receptors, which has long been purely the domain of plant physiology, has made great progress since it turned resolutely toward biotechnology. The research undertaken on synthesis of *plant fibres and lignin* is also largely based on biotechnology, as we will see in the section on bio-industries. We will also discuss the vast domain of secondary metabolism when we talk of the fields of interest to the health sector.

3.1.5. Resistance and plant protection

a) An important problem for agriculture

Protection of crops against bad weather and parasites is an important problem in agriculture at any time. In this context there are regional networks of warning, information, and advice (department or regional centres for plant protection), and the easy recourse to products offered by the powerful fertilizer and pesticide industry. The global turnover of the phytosanitary industry was close to 30,000 million Euros a year in 2001. Although these products are highly efficient, they are nevertheless inadequate and very costly for farmers as well as for the environment. Researchers therefore very early on turned towards genetics to remedy these problems. Research on cultivated plants resistant to various diseases or predators is not new and corresponds largely to ongoing breeding programmes. By means of sexual crosses and controlled selection pressures on descendants, conventional plant improvement programmes offered farmers a large number of varieties resistant to biotic stresses such as bacterial and fungal diseases and animal predators as well as abiotic stresses such as frost, heat, drought, and flood. However, characters could be transferred only between sexually compatible individuals, i.e., those belonging to the same species. Genetic engineering, as we know, overcame this obstacle and very soon raised hopes of simplifying the problems and rationalizing the solutions to them. The reality is not so simple, because in most cases resistance is specific and multigenic. Despite these difficulties, biotechnology seems to represent a third option, alongside treatment and selection. Let us look at some examples in this field.

b) Resistance to herbicides

Herbicides are often effective chemical products but a major difficulty arises from the need for specificity in agricultural use: the product must destroy weeds and spare the planted crops. *Total herbicides* must be distinguished from *selective herbicides*. There are often so-called systemic products, i.e., products that circulate in all the organs of the plants through their vascular systems. Two complementary strategies have been developed: increasing the sensitivity of target plants and reinforcing the resistance of plants to be spared by developing in them a system of detoxification of the herbicides used by degradation. One example of a total systemic herbicide is *glyphosate*, which intervenes in the biosynthetic pathway of shikimates because it competes with phosphoenolpyruvate or PEP at the level of an enzymatic reaction catalysed by *5-enopyruvyl-shikimate-3-phosphate synthase* or *EPSPS*. By treatment with glyphosate herbicide, the plant is deprived of its synthesis of aromatic amino acids and thus of numerous molecules indispensable to its growth or metabolism, such as IAA, lignins, coumarins, and tannins. A modified EPSPS gene of bacterial origin placed under the control of signals recognized by eukaryotic cells was introduced in tobacco cells by indirect transgenesis. After regeneration, the tobacco plants showed lower sensitivity to the glyphosate. Resistance was found to be increased when a signal sequence was added to the chloroplast, where natural and functional EPSPS of the plant normally functions. A third level of resistance to the product was obtained when the gene was placed under a promoter with an expression reinforced in the meristematic cells. At this final level of resistance and for precise doses, the total herbicide became selective for the transgenic plants. We have seen that other herbicides are selective with respect to major botanical groups. There are antidicotyledons (or *antidicots*) such as bromoxynil, which is better tolerated by Monocotyledons because the latter possess a nitrilase ensuring the detoxification of the product. A nitrilase gene taken from a bacterium was isolated and introduced in an effective chimerical composition (notably under the promoter of the small subunit of rubisco) into a variety of tobacco, which thus acquired resistance to bromoxylin. A similar strategy was used to introduce resistance to phosphinothricin (also called glufosinate) into tobacco. This resistance could be used in turn as a selection tool for tobacco plants that have undergone a new genetic engineering operation of a target gene.

c) Identification of pathogens

The first step in controlling a pathogen or predator is to identify it. Many methods of diagnosis or identification, adapted to each category of parasites, have been developed and overall are satisfying. They are nevertheless all time-consuming, painstaking, uncertain or imprecise, and costly in that they require qualified personnel. Very often there is a difficult protocol involving extraction of samples, various microscopic examinations, cultures, comparison of morphological, anatomical, biochemical, physiological, and growth parameters, serological reactions, and so on. During this time, the disease is very likely to develop. It thus seems important to reduce the time required to establish a diagnosis. Molecular biology can offer speed as well as precision in identifying pathogens. This is why plant pathology laboratories have turned to tests based on DNA technology to identify parasites, whether they are viruses, bacteria, or fungi. One example is halo blight of bean, identified generally at a later stage by the appearance of discoloured areas on the leaves and greasy patches on the pods. Samples followed by cultures on selective mediums followed by biochemical tests can be used to identify, rather late, the presence of _Pseudomonas syringae_ pathovar _phaseolicola_, the agent that causes the disease. It would be more useful to detect the presence of the parasite much earlier, when its population is small and when treatment can still be effective. The presence of the parasite can also be detected directly in the soil, even before the crop is planted.

The trials appear to meet the need for a rapid, reliable, and precise diagnosis up to the subspecies level, and one that is less costly. A routine technique was developed based on the PCR amplification technique, of a DNA sequence specific to the pathovar (Fig. 3.10). Research on specific sequences has also been carried out for other pathogenic bacteria such as _Pseudomonas solanacearum_, a pathogen of tomato, _Pseudomonas syringae_ pathovar _pisi_, a parasite of peas, or _Xanthomonas campestris_ pathovar _phaseoli_, which attacks bean crops. In all these examples, the specific sequences can be used to detect bacteria directly in the seeds, particularly for procedures of certification. The technique requires only a few hours, is highly specific, reliable, and reproducible, requires a small population density, and can be carried out on a plant in the early stages of development as well as on a soil sample, all at a low cost. Very similar techniques are now being developed for the _detection of viruses_, particularly in fruit trees (e.g., peach). Recently, a test was developed for the detection and identification of toxins in cereals (BASF agrotransfer, 2003). It consists of finding the presence or absence of the _Tri 5_ gene,

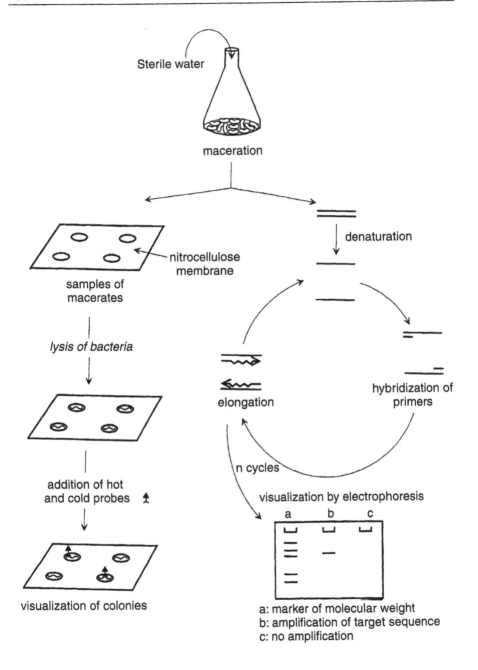

Fig. 3.10. Diagram summarizing the techniques for identification of pathogenic bacteria using probes (left) or amplification of a specific sequence using PCR (right)

which codes for the synthesis of trichodiene synthase, an enzyme that catalyses the synthesis of trichothecenes, which constitute one of the most important families of mycotoxins.

These are only a few examples and, in the future, techniques for pathogen detection by PCR will be developed in the form of kits that are accessible to the farmer.

d) Resistance of plants to viruses

Many plant diseases are due to viral infections. We have seen that one way to treat an affected plant is to regenerate it from a meristem culture. By this means we obtain a plant that is free of the disease but can still be invaded by it again. It is thus a provisional recovery and not a protection by vaccination, as an animal could be protected. We can also select varieties that are less susceptible or not susceptible to the disease. It is also possible to create resistant varieties by genetic engineering. For this purpose, a technique has been developed to isolate a fragment of viral RNA, the translation of which gives rise to a capsid protein. The reverse polymerase allows the synthesis of a corresponding cDNA. Some plants have been transformed (e.g., tobacco, potato, tomato, beet, cucumber, olive, rice) using these cDNA and it has been observed that they became resistant to the virus. The mechanisms by which this resistance is acquired are not, however, completely understood. Among the hypotheses formulated, the most plausible comes from a saturation of _decapsidation_ sites of host plant cells; decapsidation is indispensable for viruses to return to invade the cell and then replicate. These viruses, which conserve their capsid and are thus incapable of replicating, rapidly decline in number and die out. According to other hypotheses, they may no longer be recognized by cells genetically transformed by cDNA coding for capsid proteins. Resistant squashes obtained through the same process, called _pathogen-derived resistance_, were commercially produced in the United States by Asgrow in 1996. Since then, many resistant plants have been created, such as melons resistant to mosaic, plum trees resistant to Sharka virus, or grapevines resistant to court-noue virus (1990-2001). A new antiviral technique seems to have a highly promising future. It is based on the discovery, among many RNAs that the plant cell harbours, of double-stranded molecules capable of blocking certain transcription processes. This is called "epigenetic inhibition of transcription." The process is related to that known in animal cells as RNA interference mechanism or iRNA. The use of double stranded synthetic RNA molecules of around 20 nucleotides forming a loop, in which the 3' end comprises three supplementary or non-paired nucleotides (Fig. 3.11), leads to the degradation of messenger RNAs that will have the same sequence, especially those the cell uses for

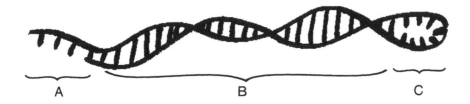

Fig. 3.11. Double stranded synthetic RNA molecules forming a loop 3 non-paired nucleotides (A), region of paired nucleotides (~20) (B) and a loop of 9 non-paired nucleotides (C).

the synthesis of capsids of viruses that infect it. This degradation involves a ATP-dependent RNase III, called "dicer".

Other techniques, chiefly involving animals, are being experimented on and the term *plantibody* is now being used, referring to virtual antivirus antibodies produced by transgenesis through plant cells. These techniques are still at the experimental stage and the feedback is not yet sufficient to certify the harmlessness of the technique (there may be recombination between the virus and the transgene, a phenomenon of complementarity restoring to the virus all its virulence). The stakes are nevertheless so high that such studies will be pursued despite the undeniable difficulties. In 2001, some of these constructions were in the clinical trial stage, but since then none of them has been authorized for marketing.

e) Resistance to insects

We have already described a strategy in which the sterol profiles of the plant are modified by transgenesis in order to influence the moulting potential of insect populations. Other effective techniques, presently well developed, have been proposed. Among these, the best known is the Bt technique. It is named after *Bacillus thuringiensis*, known for its property of secreting insecticidal toxins. In fact, these toxins modify the permeability of the membrane of intestinal cells of the insect in its larval form (caterpillars, for example), profoundly disturbing its nutrition and ending in the direct absence of moulting and, very often, the death of the insect. Suspensions of these bacteria have for a long time been used as biopesticides on crops when the threat is identified. The technique is, however, relatively costly and effective only for insects that live on the leaf surface. It is much less effective for insects that dig galleries in the plant, such as maize pyralid. To resolve this latter problem, biotechnologists proposed to introduce the gene coding for the synthesis of the toxin directly into the genome of the host plant. Some encouraging preliminary results were obtained with the terminal part of the native gene, under control of the 35S promoter, in tobacco and tomato plants. Other transformations were carried out on a very large scale in the United States with the synthetic gene in cotton. The results were highly significant in the first year, but less so in the second, which raises some questions about the future of this strategy, especially following the possible development of resistance by the insect. There is, however, a time lag (7 years of observation in the United States in 2003) before we can expect a regional long-term eradication of pest populations even if 16% of the cotton plants on average maintain a resistance to the toxin. Some researchers recommend that observations be made over about 10 years.

Researchers today have available a variety of Bt genes adapted to various populations of phytophagous insects. The largest application, and the best known at present, is the creation of transgenic maize varieties expressing Bt genes. This technique was widely developed in the United States since 1996 but met with reluctance in France. In 1998 it was authorized for consumption but not for cultivation, and in 2003 it was authorized for cultivation, provoking highly publicized debates between opponents and advocates. Among the reasons for opposition, the most frequently mentioned is the mode of selection of plants by ampicillin, even though this last character is not ultimately found in the transformed plants. Others, such as the risk of dispersal of transgenes through crosses, are less plausible, given the sexual isolation of maize in Europe. There are also some variant techniques using genes coding for the synthesis of proteins with insecticidal activity (e.g., protease inhibitors or cholesterol oxidase gene).

An elegant technique that has not been widely exploited is the production, by the transformed plant, of molecules that can attract predators of phytophagous insects. This technique is based on a natural process identified in the years 1997-2000. The molecules, for example linalool, belong to the terpenoid group, which appear as chemical mediators intervening indirectly in the plant's defence mechanisms against pests.

No matter what the strategy used or debates generated, techniques of genetic engineering for crop protection against predatory insects are very likely to develop further in the near future. For example, more than 50% of the cotton produced in Eldorado (USA) expresses the cry 1Ac toxin of Bt.

f) Resistance to Mycophytes

The other major category of parasites of cultivated plants is represented by many species of fungi. Although they are still essentially controlled by antifungal treatments, biotechnological measures have begun to be validated. One of the most effective approaches seems to be the introduction into the plant of genes coding for the synthesis of enzymatic proteins with antifungal activity, such as ß-glucanases or chitinases, which, alone or together, attack the cell wall structures of the fungal cells. An acceptable level of resistance often requires the presence of several transgenes. A second approach is the development or reinforcement of protection by introducing a gene favouring the biosynthesis of an antifungal compound of the phytoalexin type. In this way, the gene for stilbene synthase, extracted from grapevine, was transferred to tobacco to make it resistant to rot. A third approach is to introduce into the plant a gene coding for the synthesis of an inhibitor of an enzyme indispensable to its infectious capacity, such as polygalacturonase. These different approaches can be combined to achieve a certain level of effectiveness. So far, versatility of protection has been more sought after than specificity. The case may be completely different when the antifungal control is aimed at a particular parasite. In that case, genetic engineering is primarily used in precise identification of the pathogen. Various techniques of identification by molecular biology have been proposed to replace conventional isolation followed by culture on a gel medium. Using specific nucleic probes for a particular gene sequence of pathogenicity of a fungus, we can detect, as we have seen for bacteria, with equally great specificity, sensitivity, and rapidity, the presence of the pathogen on the crop or, what is much more useful, its presence in the soil even before the crop is planted. A list of transgenic plants resistant to pathogens is given in Fig. 3.12.

Virus	Bacteria	Fungi	Insects
Tobacco	Tomato	Tobacco	Tobacco
Potato	Potato	Tomato	Maize
Rice	Tobacco	Rape	Tomato
Maize	Rice	Carrot	Rice
Soybean		Kiwi	Peas
Melon			Sugarcane
Citruses			
Peas			

Fig. 3.12. Some transgenic plants resistant to pathogens.

g) Other resistances

Many other types of resistance acquired by biotechnological means have been studied. Among them, plant resistance to drought and salinity, tolerance of water stress, and adaptation to unusual environmental conditions have been subjects of research in some laboratories in Mediterranean countries. In 1999, a tomato plant was created that had received the gene for PEP carboxylase of sorghum and was said to be resistant to salinity and very economical in use of water. These studies used models, and control of the synthesis of abscisic acid by the plant seemed to be one of the keys to the problem. However, drought resistance seemed in all cases incompatible with the best parameters of growth and yield.

3.1.6. Molecular markers and plant breeding

A molecular marker is a DNA fragment, generally small (200 to 2000 bp), the sequence of which differs from one plant to another or from one species to another, even in plants that are otherwise very similar. These differences manifest themselves in the presence or absence of a band, band size, or interval of the band on electrophoretic profiles. From these observations it is possible to establish "genetic maps" of a given species taking into account the number of markers, which could be extraordinarily high. The markers are present regardless of the physiological or environmental conditions of the plant. These maps can subsequently serve as the basis of "marker-based breeding programmes". The task consists of a preliminary step of locating the genes of interest in relation to these markers, keeping in mind that the closer they are to one another, the greater the chance that they will be transmitted together in their descendants because of linkages. The advantage is particularly great for genes that have a heterozygous stage since the search for a marker in a DNA extract is infinitely quicker than the search for a recessive gene by genetic techniques of crosses. By this means, the heterozygous plants that are derived from a cross can be found very quickly. These technologies are particularly valuable in following the heredity of male sterility genes used in plant improvement. They also make it possible to control varietal purity in seed lots.

Molecular markers applied to selection and the creation of varieties are so useful that all breeders use them. If the organism or seed group is large enough, a specialized laboratory is devoted to this technology (e.g., Biogema for the Limagrain Company in France). The task can thus be sub-contracted to smaller organizations that specialize in it. Agrogene, located near Paris, studies all the problems of molecular markers involved in plant improvement programmes. Molecular markers are also increasingly used in varietal identification for the purpose of property and certification controls. Control organizations, such as GEVES, have in the past few years moved from varietal identification by marking proteins (electrophoresis of isozymes on starch gel) to molecular biology (RFLP, RAPD, mini- and microsatellites).

Molecular markers can also be used to reconstruct the origin of a species.

3.1.7. A provisional assessment

Plant biotechnology has certainly infiltrated all the conventional sectors of agronomy. It has begun to have an impact on agriculture and represents a considerable advance in that domain. And the technologies have just begun. Although some transgenic crops now cover millions of hectares, especially in the United States, these GMOs and their derivatives are still present in only a minimal part of our food. This will not continue for long and the future is shaped largely by the research undertaken and the decisions made in the present. The agricultural profession is largely convinced of the economic advantage of these biotechnologies and the awareness of the stakes they represent in food production on the global scale, involving billions of Euros. In addition to the modification of agronomic characteristics of plants themselves, the processes of transformation of agricultural products to food or industrial uses are being modified and these sectors of the economy are likely to find a better return on investment when they use such technologies. The importance of plant biotechnologies is evident from studies and results, as has been remarkably expressed by H.S. Chawla in his comprehensive and clearly written *Introduction to Plant Biotechnology*. This future will also involve acute awareness of the potential risks, greater caution, and stricter observation of rules that consumers, that is to say, each of us, will impose.

3.2. PLANT BIOTECHNOLOGY AND BIO-INDUSTRIES

3.2.1. A quick overview of bio-industries

Bio-industries did not arise after the emergence of biological engineering. They have in fact existed since the dawn of time. The production of wine by the ancient Egyptians and the fermentation of beer by the Celts were already perfect examples of transformation by living organisms of material coming from the metabolic activity of other living organisms. From the 19th century onward, these activities were no longer cottage industries but took on an industrial dimension, along with other sectors of the economy.

The impact of biotechnology in general and plant biotechnology in particular in the bio-industrial sector is still quite limited. Nevertheless, it grows continually and we can predict, at least in some sectors, considerable progress of these technologies in the overall economy.

At present, bio-industries can be defined as activities carried out chiefly in two major sectors of economic life: the *agrofood* sector and those linked to *health*. They are also involved, with varying impact, in other sectors such as *environment, raw material* (especially wood and paper pulp), and, to a small extent, *energy*.

Bio-industrial organizations were at first small and highly specialized structures. They were often financially supported or owned by large groups belonging traditionally to the sectors of energy or chemistry (for example, Elf, Rhone-Poulenc, ICI-Zeneca, Roussel-Uclaf, and Hofmann-Laroche in the past, and more recently Aventis, Adventa, Syngeta, Pioneer, and Monsanto).

Still, it is not rare to see some large-scale bio-industrial structures extend their activities into two or three sectors simultaneously, often under different trade names but belonging in fact to the same holding. The borders of these sectors are not always easy to define, since they are constantly altered by sale, acquisition, reorganization, merger, transfer, and division. All these changes result in a highly complex network of alliances and markets, very difficult for most people to follow.

The bio-industrial sector has moreover given rise to a multitude of small companies specializing in the construction and sale of laboratory equipment or supplies. These products are valuable aids to research and the catalogues published by the companies are mines of information on molecular biology and biotechnology. This sector is no longer safe from restructuring and reorganization. Bio-industries have also contributed to the emergence of highly specialized *service organizations* working in public research laboratories as well as private ones to provide genome analysis, oligonucleotide synthesis, sequencing, kits, and other products and services.

Biotechnologies have advanced most in the bio-industries using microorganisms: bacteria and yeast. Fermenters and bioreactors using reprogrammed microorganisms function at all scales: experimental, pilot, and industrial. The vast sector of health benefits most from this domain and ethical dilemmas do not seem to have been an obstacle. It is very different in the case of agrofood, where the reservations and restrictions are much more numerous and accentuated. We will address some aspects of these problems in the next chapter.

In the industrial sector, biotechnologies using cells of higher organisms—plants and animals—are still, with some exceptions, searching for credibility. They are nevertheless part of many programmes, often as a complementary or alternative route. With respect to molecules with a potentially high added value, the choice lies between extraction and biosynthesis. The solution adopted is generally determined by economics, more rarely by ethical considerations, and exceptionally by both.

Studies in plant biotechnology in the context of economic programmes are helped by a number of coordinating or financing organizations in Europe. These programmes were initiated in 1982 and have gradually linked fundamental research to applications, creating platforms and initiating partnerships between public laboratories and industrial groups. There was certainly some meandering, cacophony, and overlapping between programmes in the 1980s, but major entities ultimately emerged with some common objectives necessary for basic knowledge such as the genome sequencing of *Arabidopsis*, a model, albeit inedible, plant (Fig. 3.13). These joint programmes have had the advantage of considering the plant as a "molecule factor" that can be used for food or non-food purposes. They certainly have some limitations, since they concentrate on one family of products at the expense of another (e.g., fructose polymers), but they are nonetheless a powerful driving force for biotechnological research in the projects they generate.

Fig. 3.13. *Arabidopsis*: a model plant. Long considered a weed, *Arabidopsis thaliana* has become a model in plant molecular genetics. It belongs to the family Brassicaceae, earlier called Cruciferae. It grows to a few tens of centimetres and can complete its vegetative and reproductive cycle in less than two months. Five to six generations in a year can therefore be followed genetically. Its genome, distributed over 10 chromosomes in the diploid state, is the smallest genome known among plants and only seven times as large as the yeast genome. This is why it was chosen for an integrated sequencing programme. A large number of mutants are presently known, affected in physiological functions or the development of certain organs, particularly the flower. The genome of *Arabidopsis*, which contains at least 25,000 genes, was almost entirely decoded and published in December 2000.

We will now look at some sectors in which biotechnology applied to plants has, or will have, an impact on the direction of research and activity.

3.2.2. Agro-food industry

The agro-food industry lies directly downstream of agricultural production. It represents, in some regions, up to 18% of employment. The links it has with agronomy are so obvious that it is often difficult to separate the two activities. One centres on production, the other is an industry of transformation and conditioning for the purpose of animal and human nutrition. What they have in common is the plant and mastery of it. Food comes essentially from the plant and the quality of our food is a fundamental concern of our time throughout the world. "What will we eat today?" is a question that, even recently, was essentially *quantitative*, while today in Europe it has a more *qualitative* connotation. The Food 2000-2002 programmes initiated by the Ministry for Agriculture and Fisheries considered both aspects and it is hard to imagine that most of the foods we will consume in the future do not exist today. Food is, however, a faithful reflection of civilization.

The agro-food industry is still closely linked, in its processes as well as its image, to cottage industries, with its contradictions between the need for consistent quality and supply and the instability and variability of production that characterizes farming. Biotechnology, even though it is prominent in this sector, has not yet revolutionized either the techniques of production or the habits of consumers.

Food products are complex and difficult to classify. To take some examples, we will consider the dominant element according to conventional biochemical distinctions between sugars, lipids, proteins, and the wide variety of other nutritional products, orangoleptics or, more simply, sweeteners.

a) Glucides

Sugars and glucides represent a considerable part of the agro-food sector (Fig. 3.14). The most important representatives of this sector are commercial sugar or saccharose, saccharose polymers such as starch (the principal constituent of flours), and fruit pectins. There are sugars of lesser economic importance, such as fructose polymers, which have some units to some tens of carbon atoms, that are important to several sectors—including pharmacy and agro-food economy—because they are considered low-calorie sugars. They are produced in small quantities by plants and the species raised for this purpose are rare. Some Liliaceae, Poaceae, and Asteraceae are known for being edible in the raw state as they contain larger quantities of these polymers than others (garlic,

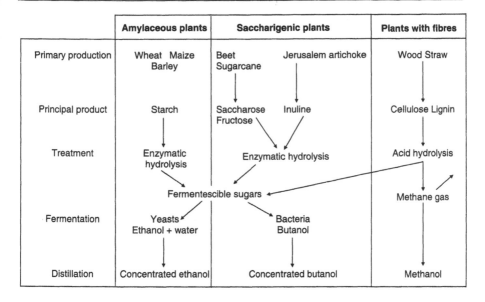

	Amylaceous plants	Saccharigenic plants		Plants with fibres
Primary production	Wheat Maize Barley	Beet Sugarcane	Jerusalem artichoke	Wood Straw
Principal product	Starch	Saccharose Fructose	Inuline	Cellulose Lignin
Treatment	Enzymatic hydrolysis	Enzymatic hydrolysis		Acid hydrolysis
	Fermentescible sugars			Methane gas
Fermentation	Yeasts Ethanol + water	Bacteria Butanol		
Distillation	Concentrated ethanol	Concentrated butanol		Methanol

Fig. 3.14. Sugar networks. A summary of the various bio-industrial pathways from sugar polymers—starch, soluble sugars, and cellulose— for production of energy content.

onion, artichoke, endive, sunflower). Among these sugars, we should mention inuline from Jerusalem artichoke, which can crystallize and the molecule of which is formed of small polymers of less than ten carbon atoms. These sugars are considered "low-calorie" because, even though they have an agreeably sweet taste, they have a poor ability to get hydrolysed during digestion and consequently supply very few calories. They also play a considerable role in the supply of fibres needed for good intestinal transit. They can nevertheless be hydrolysed later by bacteria living in the digestive tract. These sugars can also be used as emulsifiers to replace oils in some preparations. Even though they can in some cases give rise to fatty acids, they will have a beneficial effect in the regulation of cholesterol.

Genetic transformation of endive and sugar beet has been attempted to produce fructose polymers of controlled length, particularly inuline. The difficulty lies in the need for a double enzyme reaction to obtain a chain, but a system using this principle in endive has functioned relatively well in the laboratory and on a pilot scale. Genes of the fructosyltransferase type, involved in the synthesis of highly branched polymers, have been borrowed from bacteria and have experimentally been transferred with considerable success into tobacco, beet, potato, and maize.

Plants have been genetically transformed by genes coding for heat-stable enzymes that could control the constitution of sugars such as α-amylase, ß-glucanase, and xylanase. These enzymes are then stored in the seeds and can be extracted easily from them, then used practically in purification. Such processes have been applied to animal feed to improve its digestibility. Genes coding for the synthesis of these enzymes have also been introduced into barley to improve beer production processes.

b) Lipids

In the domain of lipids, which can be considered highly economically sensitive because it is controversial, the products are many but little known because they are often protected by patents. This relative secrecy of results pertains to the genes as well as the innovative processes. However, it is known that the objectives are modification or increase in production of some categories of lipids important for the lipochemistry and energy sector. As in other fields, the strategies developed are overexpression or suppression of enzymatic activities following gene transfer under a suitable promoter. The impact could be felt in the agro-food as well as bio-industrial sectors; the frontier between these two is often purely conventional. For example, after having searched relentlessly for the presence of erucic acid in rape in the 1960s and 1970s by means of multiple pedigree selection, scientists turned to biotechnology to find compatible levels of erucic acid because it has numerous uses in the polymer industry. This sector is actually a high consumer of biodegradable plastics such as polyhydroxyalkanoates. Bacterial genes controlling the biosynthesis have been transferred into *Arabidopsis* and have allowed the initiation of this activity, which is normally entirely absent in plants. Similarly, introduction of a lipase gene expressed in the seed has been attempted by genetic engineering to mobilize lipid reserves precociously.

c) Proteins

Even though all proteins have a distant plant origin, for human beings they are usually sourced directly from animals. In fact, the animal is only a link in the chain of transformation and must undergo a supplementary phase of industrial transformation before consumption (salting). Even though the per capita consumption of meat is highest in France, the proportion of meat in the diet has fallen for the past few years. The meat consumed in the greatest quantities worldwide is pork, which represents 30 to 45% of the total meat consumption. The intervention of biologists in the domain of meat production involves, on the one hand, the mapping of the genome of the species most commonly consumed for applications in breeding (e.g., establishment of the pig genome map) and, on the other hand, food security and microbiological control of food (research on *Salmonella, Listeria, Clostridium*, research on a bacterial flora capable of destroying these germs, such as anti-*Listeria* bacteria). With respect to meat itself, the results of genetic engineering have not yet reached genetic engineering, since the image of a trademark could be tarnished by these applications, even though some people acknowledge the advantages of transferring a gene for resistance to some diseases.

Plant proteins that enter into the composition of cattle feed come either from the aerial part of the plant after dehydration (e.g., lucerne proteins) or fermentation (ensilage), or from proteins accumulated in the seeds of proteaginous species. At present, meat is essentially improved through enhancement of the digestibility and palatability of plants fed to cattle. For this purpose, assays in the phytase enrichment of forage were undertaken to improve the assimilation of phosphorus by consumer animals. The biotechnological stakes in this case are related to the problems of improvement of forage plants and cultural practices that were envisaged in agronomy.

d) Aromas and perfumes

Aromas constitute a vast category of molecules that are mostly extracted from microorganisms, plants, or fungi. There are a certain number of so-called artificial aromas, i.e., in principle obtained by synthesis but having to be identical to the natural aromas in order to be used in agro-food. This is at least the legislator's aim. Today, there is a much richer palette of artificial aromas than of aromas actually extracted from living organisms. The biotechnological means seems to be preferred to obtain synthetic products that are very similar to natural products, even though the consumer is still actively concerned about having a "natural" product. This biotechnological route leads to syntheses in which enzymes, microorganisms, and plant cells intervene in processes of fermentation or bioconversion. The return on investment is also a critical aspect of the choice of technique and products of biotechnology have been found to be more profitable than natural extracts. In chemical terms, aromas are often *short chain fatty acids, aldehydes* (resulting from the metabolism of amino acids), *secondary alcohols, esters, lactones, methylcetones* (from fatty acids) and *terpenes* (derivatives of mevalonic acid). In nature, there are often complex combinations of these various groups of molecules. These aromas are presently widespread in numerous agro-food sectors: they are found in, for example, fermented milk products, meat, drinks, preserves, baked items, and pastries.

Genetic modifications have been introduced in microorganisms such as lactic bacteria and yeasts as well as in filamentous fungi and plant cells so as to make them produce substances, either because such substances are not synthesized naturally or because they are synthesized in very small quantities. Industries prefer to use microorganisms, while the public prefers plant cells. There are presently about a hundred aromatic molecules on the market that are obtained by biotechnology, a good part of which come from the activity of yeasts with modified genetic material. Some yeasts have been programmed for the bioconversion of alcohols and aldehydes in their corresponding acids, a bioconversion that is accompanied by a change from an acid taste to a sweet and fruity taste. Other yeasts, blocked in the sterol pathway, secrete geraniol and linalol, which are dominant aromas of muscat scent in wines. The same is true for menthol, which can be synthesized from a substrate of polyunsaturated fatty acids and soybean flour, which provides the enzymes needed for bioconversion. Some carotenes can be transformed into ionones. Raspberry scents are extracted by bioconversion of molecules present in birch. Lactones are widely used for the manufacture of many fruit scents. The same is true for the production of monoterpenes from geranyl diphosphate. Vanilla and vanillin, its biosynthetic substitute, are other examples of natural aromas, while ethylvanillin, which has a more pronounced taste but is not present in the vanilla pod, is considered an artificial scent.

The difference between the natural perfume and its artificial homologue is sometimes minimal and fraudulent uses can be detected only through detailed analysis, such as nuclear magnetic resonance associated with mass spectrometry. In France, Pernod-Ricard is the leading organization in this field. In the United States, Senomyx and other companies operate entirely on the genomics of aroma and perfumes designed for cosmetics and agrofood.

Strategies for overexpression to increase the biosynthesis of a useful metabolite have frequently been used, but tools that can act simultaneously on all stages of the biosynthesis pathway must also be used. This is why research has been directed toward the control of regulator genes (e.g., str or t genes of ORCA 3)

In the field of perfumes and aromas, we can expect a considerable growth of biotechnologies.

3.2.3. Fibre industry

For some years, the development of plant production for non-food purposes, i.e., industrial uses for which GMOs can be used without difficulty, has been encouraged. This is already the case with transgenic cotton cultivation in the United States. These crops can be grown on fallow land in Europe.

The focus is mainly on fibres of plant origin such as wood and its derivatives, cardboard and paper. Lignin, which is found essentially in trees and other woody plants, and cellulose are the major constituents of vascular cell walls specialized in the conduction of sap and the support of the plant. Lignin and cellulose together represent nearly half of the biomass on the earth's surface. These molecules are highly polymerized, make the walls waterproof, and are relatively difficult to hydrolyse, which is why they are effective against predators and parasites. They are responsible for the rigidity of plants, their general habit, and their resistance to heavy winds.

Celluloses and hemicelluloses are glucose polymers, while lignin is derived from phenylalanine through a biosynthetic complex. The enzymes that catalyse the steps and a certain number of genes that code for their biosynthesis are presently known. From this knowledge arise various projects of genetic engineering designed to control or modulate lignin composition in plants. Lignins, we have seen, are unfavourable to the digestibility of forage by ruminants as well as to the manufacture of paper. Thus, they must be extracted before pulping but their solubility is highly variable depending on the species and even depending on the plant's stage of development. Lignins are often difficult, in some cases practically impossible, to extract. This specific variability may serve as a model or basis to attempt to reduce an enzyme that intervenes in the last step of lignin synthesis by genetic engineering. This is done by placing in antisense the gene that codes for the synthesis of CAD (cinnamyl alcohol dehydrogenase). Modified tobacco present a very slightly reduced level of lignin but its chemical structure is profoundly modified, without any notable disturbance in the normal characteristics of the plant. Such modified lignins can more easily be extracted chemically. Consequently, fewer bleaching products are used for paper pulp and the process is safer for the environment. Antisenses of the gene for CCR (cinnamoyl-CoA reductase) have also been used in genetic transformation and have led to a reduction in lignin levels and increased extractability. However, anomalies in growth have sometimes been observed in these cases. The use of the gene for OMT (O-methyl-transferase) in antisense construction reduces the degree of methylation of lignins by replacing the $O-CH_3$ by hydroxyls and thus profoundly affects the nature of products, making them more favourable to digestibility than to industrial uses. Double transformants seem to have accumulated several advantages, especially in poplar, a tree of great importance to the paper industry. Similar programmes exist for spruce and eucalyptus. The impact of such techniques can easily be imagined when it is known that the paper pulp industry represents the third-ranking item of deficit in the economies of countries of the European Community.

3.2.4. The health sector

It is well known that most drugs are derived from plants or plant extracts. These close relationships have not been weakened by the advent of biotechnology. The search for new species of botanical varieties is today motivated as much by the acquisition of fundamental knowledge as by practical pharmaceutical uses of newly discovered natural substances that such plants may contain and eventually produce. The major public laboratories (the CNRS or ORSTOM laboratories of natural substances) and private laboratories (Zeneca, Merck, Rhone-Poulenc, Boiron) have their teams of *botanists, pharmacobotanists*, and *ethnobotanists*, who travel through those regions of the planet (such as the Malagasy forest or Amazonian forest) that have been less exploited for these purposes. They use small robots capable of detecting the presence of natural products suspected of having enzymatic and pharmacodynamic properties, and thousands of samples are treated and assayed each week. Such exploration can develop further, given that less than 5% of the flora has been studied from a pharmacological angle. However, it is estimated today that there are 13,000 plant species used in pharmacy, given that 75% of active ingredients contained in medicines are derived from plants. These advances have paid off these past few years with the discovery of molecules such as cyclosporins, vinblastin, vincristin, taxol, and taxotera. The reduction of biodiversity by the uncontrolled exploitation of forests and its regrettable impact on the potentialities of pharmacy could be considered one of the greatest tragedies of our time. Research in this field, while it is still conventional in its approaches, certainly has a promising future.

Biotechnologies have also been introduced in the context of enrichment of pharmacopoeia. Involving essentially microorganisms at first, they have increasingly involved plants since techniques of genetic engineering of plants have developed and have made the plant a potential factor for the production of recombinant proteins for the pharmaceutical industry. Plant cell culture offers several advantages, among which are the capacity to produce pharmacodynamic molecules in large quantities under highly competitive economic conditions. A second advantage arises, for proteins, from a capacity for post-translation maturation such as glycosylation, secretion, cleavage of the peptide signal, folding, and formation of disulphur bonds that generally occur in mammal cells. The first transformations of plant cells for industrial purposes were carried out for heat-stable enzymes for a food purpose. At present these transformations mostly have a therapeutic objective. Plants have, in addition to their great aptitude for transformation, the advantage of not having to eliminate oncogenic

DNA or pathogenic viruses used for the transfection of animal cells. Plant viruses are, in effect, inoperative on human cells.

The production of drugs by plants is no longer a project but a reality. For example, mice have been vaccinated against endotoxin of *E. coli* by consuming plants (transgenic potato and tobacco) expressing a subunit of the toxin. The mice then secrete the antibody and are immunized. The same protection could be given to humans because oral vaccines through fruits consumed raw, especially in banana and potato (for hepatitis B), have already been produced. These technologies open up immense perspectives in developing countries, which often do not have the means to use the same vaccination techniques as the industrialized countries and could also find a new agricultural market.

Therapeutic molecules of primary importance could be or are already produced by reprogrammed plants. Examples are gastric lipase, the growth factor of epidermal cells, γ interferon, serum albumin, human encephalin and haemoglobin, antitumorals such as vinblastin, vindolin, and novelbin, and hypotensives such as ajmalicin. In this last case, the plant system (tobacco and potato) is found to be more effective than *E. coli* and even yeasts, since the molecule matures in perfect conformity with that realized within human hepatic cells. This conformity is, however, difficult to obtain in some cases, because the later steps of the maturation occur in highly specific intracellular compartments. It is therefore necessary, at the level of the recombinant protein, to predict the correct target signals and, if possible, induction signals in order for the synthesis of the product to be triggered in the plant at a precise stage of development, for example before harvest. It is also sometimes necessary to arrest the maturation before the plant cell imprints its specificity and consequently alters the biological activity of the synthesized molecule. Some studies have been undertaken on the modifications of the glycane synthesis pathway so that the plant realizes a glycosylation of the protein that will be identical to that realized by mammal cells.

Economists have even calculated that the cost of production of albumin through transgenic plants could be one fifth that of the albumin obtained by purification from blood plasma. The global needs for human albumin could be covered by cultivation over just a few thousands of hectares, or even less if particularly effective varieties for this type of metabolism are selected.

Using similar processes, other proteins of biological interest, such as serum albumin, acid lipase, and hirudine, have been synthesized. Human serum albumin has been produced by transgenic tobacco and potato, under promoter 35S. The proteins were isolated and purified from leaves or cell cultures, but the production capacity of the plant was found to be relatively low. Even though the process was technically demonstrated, the expression of the proteins must be optimized if it is to be commercially viable. The acid lipase, which has an optimal enzymatic activity at pH 4.5 and resists the proteolytic properties of gastric juices, can be used to treat persons affected by mucoviscidosis, who because of a pancreatic deficiency have difficulty in absorbing fat bodies. The gene for gastric lipase of dog was cloned and transferred into tobacco and rape; it was correctly expressed in the plants and the enzyme could be extracted and purified. In France, Meristem Therapeutics produced a mammal lipase through a transgenic maize (2000). Hirudine is an anticoagulant that is found naturally in the salivary secretions of leeches. The gene was cloned and transferred into rape under a promoter leading to the accumulation of a recombinant protein associating hirudine with a highly hydrophobic protein in the seeds. The anticoagulant was then extracted and purified and proved to be perfectly active.

The most spectacular result of our times is undoubtedly the production of human haemoglobin by transgenic tobacco despite the relative complexity of this tetrameric molecule (2 α globins and 2 β globins and haem). Thus, apart from the synthesis of subunits of the molecule, correct assembly is needed to obtain a functional recombinant haemoglobin. The subunits were produced under promoter 35S with targeting to the chloroplastic compartment, where they are assembled with haem molecules and become functional, as has been shown in physiological tests of oxygen transporter. The plant cell was already shown to be perfectly manageable and suitable for these complex syntheses. It has also been shown to be so in the synthesis of immunoglobulins, multimeric proteins with their heavy and light chains and their protective compound of proteolysis. The operation required the obtaining of four lines of transgenic tobacco, each having one of the four transgenes coding for the synthesis of one of the elements. The lines were then successively crossed till a functional recombinant immunoglobulin was obtained.

Plants have numerous advantages over other transgenic systems, particularly the safety and conformity of the product synthesized. This technology allows us to overcome any threat from infectious agents since no such agent is common to the plant and animal kingdom and the processes of maturation of plant cells are largely identical; the recombinant proteins produced are very often physiologically active. In addition, there are advantageous economic conditions and a useful contribution to the policy of diversification of agricultural production. Plants have always been at the heart of pharmacopoeia, and their position is now reinforced by biotechnology.

There is one small sector of health that must not be omitted in which some plants play a role: biomaterials. Biomaterials used in medicine are polysaccharides of plant origin that have been tested for the treatment of burns (corn starch), the effective conditioning of medicines, especially for their gradual release and targeting in the patient's body, and the manufacture of dressings (alginate of algae). Pectins are currently the subject of many studies. For example, *bupleuran* 211b is a factor of lymphocyte B differentiation and 211c is used to treat certain auto-immune diseases.

Drugs of biotechnological origin are protected by patents, which makes them particularly costly. Nevertheless, some of these patents will soon lapse into the public domain (mainly in 2005-2006) and many laboratories will begin commercial production of generic biotechnolo-gical drugs, which will be much cheaper.

3.2.5. The environment

For many decades, the environment was the concern of naturalists (ecologists, botanists, zoologists) rather than the man on the street, much less the politician. This is no longer true today and some economists predict a brilliant future for technologies for the preservation of the environment, either preventive technologies that avoid the use of pollutants or remedial technologies of decontamination. The major types of pollution are chemical (e.g., metals, hydrocarbons, pyralene) or biological (e.g., waste water, pig manure, nitrogen effluents). There are chemical remedies (solvents, dispersants) and biological remedies. Biological decontaminants can include bacteria, fungi, and plants.

A great deal is expected from plants either as pollution indicators or directly as decontaminants, especially of soil.

Many plants are sensitive to the presence of atmospheric pollutants. They could be incorporated in systems of biovigilance. It is estimated that several hundreds of plants can serve as biocaptors but in fact a much smaller number are presently used, either as *bioindicators* or as *bioconcentrators* for subsequent chemical detection. From the observation of necrosis on certain test plants, some pollutants can be detected and eventually roughly quantified. Some kinds of tobacco can indicate the presence of ozone from combustible gases. They can accumulate large quantities of pollutants without suffering much. Rye-grass used for lawns, spinach, clover, tulip, plantain, and water hyacinth can accumulate heavy metals such as cadmium, lead, mercury, zinc, iron, and nickel and can be useful for chemical measurements.

The presence or absence of certain varieties of plants also reveals a level of pollution. Lichens are good examples of such plants. This is called *biomonitoring*, especially to evaluate sulphur dioxide or chloride pollution. Test plants such as tobacco or *Arabidopsis* can indicate radioactive pollution in a simple manner. The appearance of blue patches on their leaves reveals the frequency of cells that have undergone homologous recombination under the influence of gamma rays. Stations using biocaptors are reliable and cheap and in some landscapes allow the establishment of precise maps of pollution in a given region.

Some processes of decontamination can be accelerated by the use of fungi, particularly for biodegradation of lignin, polycyclic aromatic hydrocarbons, and chlorine derivatives. The degradation process involves hydroxylation. Some plants have also been used, and this is called phytoremediation. These plants can fix undesirable elements by preventing their migration in the soil. They can also extract them and then distribute them in less toxic methylated forms. Or they can simply accumulate them. In this case, harvest followed by incineration can "clean" the soil. The ashes can be reused in metallurgy. The efficiency of the plant is linked to the development of its root system as well as that of the soil microflora and microfauna that live in symbiosis with the roots. Another parameter of efficiency relates to the mobility of the metal in the soil, a mobility that is itself linked to the soil pH. Among the plants that fix heavy metals is *Thlaspi caerulescens*, capable of extracting 30 to 125 kg of zinc per hectare. Some species of *Alyssum* are also highly effective in concentrating nickel. These plants are capable of neutralizing the toxicity of the metal by complexing or chelating it and then storing the complex in vacuoles, where it is sequestered. *Alyssum* species chelate nickel by the intermediary of an amino acid, histidine. Other plants use citric acid to do the same thing. Even on soils that are mildly contaminated, these plants end up accumulating large quantities. The reason for this metallic hyperaccumulation by some plants is not known, but some authors have suggested that it is a means of protection against parasites and predators.

Soil decontamination is not the sole use of these plants. They also offer an economical way of extracting metals from soils when the ore content is too poor for extraction by traditional means. For example, a tree of Oceania, *Sebertia*, accumulates up to 20% of nickel in its latex. Some plants have been found to accumulate lead (*Armeria, Thlaspi*), cobalt (*Silene*), or selenium (*Astragalus*).

Biotechnology has helped improve the capacity of root systems to extract metals and transfer them towards the aerial organs. In this way, the growth of accumulating plants can easily be improved, as well as their morphology, reproduction, and adaptation to new climates. Studies using classical selection, protoplast fusion, and genetic engineering have been carried out in Brassicaceae in order to develop phytoextraction of other elements, such as mercury, caesium, strontium, uranium, copper, and arsenic. Encouraging results have been recorded with *Arabidopsis*, transformed by means of transgenes of bacterial origin. Other projects involve rape, especially for the extraction of selenium. There is a joint research programme on the all-Europe level (Phytorem programme).

Phytoextraction for decontamination and extraction of elements is a new sector of economic activity that is relatively welcome to people.

3.2.6. The energy sector

The common agricultural policy of the European countries has led to a vigorous expansion of farms directed toward the agrofood sector to the point where surpluses have occurred, the management and storage of which is a serious burden on the economies of these countries. However, it must be borne in mind that these "surpluses" are localized and arise out of certain economic conditions. On the global scale, there are problems of uneven distribution of wealth and the demand for food remains high. The perception that there are surpluses has led to a policy encouraging fallow land that can be exploited only for industrial uses, particularly energy. This was the beginning of farm projects for the production of *biofuels*, or substitute fuels that could come from sugars (alcohols, carburol, or ETBE—ethyl tertio butyl ether) or fatty acids (diesters). The plants that are the basis of these programmes are specialized crops with high photosynthetic yields and a cost that could compete with that of traditional fuels. The projects are technically developed but the networks are yet to be established because of fluctuations in the prices of fossil fuels, which are themselves closely linked to international political relations.

Since this is a subject that is highly sensitive with regard to diplomacy and national strategies of self-sufficiency, studies are generally confidential or protected by patents. It is from the examination of patents and their geographic extent that more information can be gained on the status of studies and national development programmes. In this domain it is very difficult to separate what falls within the province of chemistry from what falls within the province of biotechnology. Nevertheless, it seems that in the latter category, the number of patents involving the ethanol network pathway is greater while the network of biofuels from plant oils or methyl esters seems to have reached maturity. Agriculture can also gain from these networks a valuable boost in terms of diversification. Wheat, beet, potato, and Jerusalem artichoke as well as forest wood serve as a basis for the primary network, while rape forms the second. In France, for example, the land area available for these crops is 1 to 1.5 million ha. It has been calculated that if 5% of biofuels were to be introduced in fossil fuels (5% of diester in diesel oil and 5% of ethanol in petrol), the corresponding crops would cover a million hectares. However, these products would be remunerative for consumers only if the current tax on them, which is about four-fifths of the sale price, were not collected.

For the ethanol network, the polymerized complex sugars (starch and cellulose) must first undergo a certain number of treatments to yield fermentescible simple sugars. The sugared juices are then fermented and it is here that biological processes appear, with the selection of colonies. Ethanol, which represents 6 to 7% of the liquid mass, is separated and concentrated by distillation. Combined with isobutene to give ETBE, a biofuel with a high octane index, it is then directly incorporated into a fossil fuel in proportions that may go up to 20% without posing a problem for conventional motors.

The oleaginous network involves a trituration of seeds to extract oil from them; the oil can be directly burned in diesel-type motors, but because it produces soot and odour (a strong smell of frying), it is combined with diesel oil up to 15%. While biofuels are being adapted to present-day motors, technology is being developed to adapt motors to the new fuels. However, it is the methylic ester pathway, obtained from rape oils, that is developed under the name "diester" (ester for diesel, while they are actually monoesters of oleic or linolenic acid). This fuel is now used in car parks or for urban transport. The product comes from the esterification of oil by methanol with a by-product, glycerine. In this network, the biological factor is the adaptation of varieties of rape as sources of energy.

Not all the problems posed by these new fuels have been resolved, especially the retail price, which is still two to three times that of fossil fuels. However, they represent in themselves good solutions to problems of pollution since there is no emission of sulphur and the CO_2 is recycled to some extent by photosynthesis, this last being ensured by the cultivation of rape intended for fuels of the following season.

Thus, there is significant scope for diversification in the agricultural economy, but its future is closely linked to the hazards of politics in countries that produce fossil fuels.

3.2.7. A provisional assessment

Plant biotechnologies have certainly infiltrated all the conventional sectors of agronomy and have begun to have an impact on agriculture. They are, however, only just beginning. Although some transgenic plants cover millions of hectares, especially in the United States, the presence of GMOs and their derivatives still accounts for a minimal part of our diet. This will perhaps change in the future and the immediate future is strongly conditioned by the research undertaken and decisions made today. The professional agricultural groups are largely convinced of the economic advantages of these technologies and aware of the stakes they represent with respect to food production on the global sale, which involves billions of Euros. In addition to the modification of agronomic characteristics of the plants themselves, there is the modification of processes of transformation of agricultural products for food or industrial use and these economic sectors are likely to find a proper return on the investments they make in plant biotechnologies. This future also involves an acute consciousness of the potential risks, greater caution, and stricter observance of rules imposed by consumers.

FURTHER READING

Most numbers of the journal *Biofutur*, particularly nos. 163 (January, 1997), 166 (April, 1997), 169 (June-July, 1997), and 172 (November, 1997).

Carriere, Y., Ellers-Kirk, C., Sisteron, M., Antilla, L., Whitlow, M., Dennehy, T.J., and Tabashnik, B.E. 2003. Long-term regional suppression of pink bollworm by *Bacillus thuringiensis* cotton. *Proc. Natl. Acad. Sci. USA*, 100 (4), 494-498.

Department of Life Sciences, CNRS, no. 20 (April, 1997).

Elbashir, S.M., Harborth, J., Lendeckel, W., Yalcin, A., Weber, K., and Tuschl, T. 2001. Duplexes of 21-nucleotide RNAs mediate RNA interference in cultured mammal cells. *Nature*, 411, 494-498.

Smith, J.E. 1996. *Biotechnology*, 3d ed. Cambridge University Press.

Chapter 4

Plant Biotechnologies and Bioethics

4.1. EVALUATION OF RISKS

As in many other domains of scientific progress, the development of biotechnologies has from the outset given rise high hopes and grand projects, as well as mistrust and misgivings. In effect, the objective of research in this field is to master the genome: a common possession, considered valuable, fundamental, and inalienable, that is shared by all living things and links human beings to their most distant origins, from the dawn of time. It is not surprising then that, starting from a technology, we ended up in the domain of "morality" and "religion", with the awareness that there could be consequences and factors other than simple acquisition of knowledge. It was therefore necessary to establish precautionary procedures, draw up controls, define rules, elaborate a judicial arsenal, put safeguards in place, and delimit domains of permission and prohibition. This marked the emergence of *bioethics*, with its sages and witches, commissions and colloquia, and the establishment of various administrative structures charged with codifying approaches and streamlining procedures at the national and, for several years now, at the regional level. These arrangements today seem indispensable to all of us who recognize that the regulatory commissions work with seriousness and competence, even when the researcher sometimes loses courage and patience with their complex administrative machinery.

Among the biotechnologies, genetic engineering occupies a particular place because it is more controversial, especially when human genetic material is concerned. Genetic engineering in plants is also a focus of debate since people get most of their nutrition and health care from plants and thus fear that they may unwittingly absorb the possible and unforeseeable evils that scientists may have hidden in plants. It follows that public opinion today is divided between the extremists of transgenesis, who look only for the quickest profits, and those who advocate refusal without discussion, partisans of the strictest moratoria. Nevertheless, there is a middle ground between these two extreme positions, each evolving as a function of its convictions, sensitivities, interests, information that is available to it and that can be extended, and the conclusions that can be drawn from it. The first duty of scientists in this field is to contribute the most objective information, whether technical or bibliographical, and to respect all the opinions expressed. Such an attitude is not incompatible with their behaviour as citizens.

In the context of genetic engineering and one of its most innovative applications, GMOs, the judicial framework consists of legislation in each country as well as protocols established during international colloquia for the prevention of biotechnological risks that could be linked to the existence of modified organisms for human or animal consumption. In Europe, the Cartagenea Biosecurity Protocol (19-01-2000) provided for a common definition of GMO (12-01-2001), establishment of a centre for prevention of biotechnological risks, and regulation of cross-border exchanges between the 80 member countries that signed the accords (18-02-2002) as well as from and to countries outside this community (16-10-2002). Moreover, the nations elaborated a programme for the management of the environment designed more particularly to help developing countries develop GMO legislation.

The end of the 20th century was marked by totally opposing stands between the countries of the New World and the Old World, essentially Western Europe. Whereas the United States favoured the development of transgenic crops for food, the European countries decided in 1998 on a moratorium allowing for a pause for reflection and observation before deciding on the introduction of GMO crops for human and animal food. Planned at first for a period of two years and then extended, the lifting of this moratorium today seems imminent, especially under pressure from the United States and the World Trade Organization, even though the opponents of GMOs have expressed active hostility. These opponents advance certain arguments that we will address successively, hoping to contribute, even modestly, to the clarification of one of the most active debates and concerns of our times.

4.1.1. Transmission of transgenes through food

One of the questions most often posed involves the transmission through food of a transgene, the target gene or the marker gene, into humans and insertion into human genetic material. It is curious that this concern has not crossed the minds of people who moisturize their faces with widely advertised creams containing DNA. This is a problem that was interesting to the sages of antiquity, who asked why a rabbit that regularly eats raw carrots does not turn into a carrot. It is probable that the DNA that passes through the successive enzymatic secretions of various parts of the digestive tube is not in a state to play any genetic role whatsoever in the cell that receives the products of degradation. Although in this field there is no certainty on the risk incurred in the case of ingestion of a transgene, there is presently to our knowledge no element that could support the idea that there is a risk.

The marker genes that accompany the target gene have been the focus of many criticisms because, apart from incorporating themselves in the genome along with the target gene, they are often resistant to antibiotics. Some people bring up the risk of transmission of resistance to antibiotics into human beings. Perhaps it would be advisable to recall that eukaryote cells are naturally resistant to most antibiotics, and that this essential difference is the basis of the principle of antibacterial treatments. We must add that some antibiotics used in cellular selection, such as kanamycin, are not currently used in medicine. Other fears have been expressed on the possible transmission of the character of resistance to microorganisms living in the digestive tube of humans or animals in the wake of phenomena of conjugation or translation. This possibility does not seem unreasonable and it is one of the reasons that urge regulatory commissions to demand that this resistance to antibiotics be removed, after having served its purpose of selection, from plants susceptible to experimentation in the field or designed to enter commercial production. This removal, obligatory from 2004 onward, is not always easy and therefore there are still difficulties in some research programmes. At present, there are several research programmes aimed at the suppression of all the risks linked to the presence of these marker genes either by total removal of the sequence or by slight modification of the gene present in the genetic patrimony of the plant. In other cases, as in pyralid-resistant maize, the techniques of transformation used do not allow us to find, apart from the target gene, the other genes of the construction, particularly the gene for resistance to ampicillin, chosen by the experimenters for selection of transformed cells. Only the target gene is integrated in the genome and this methodology seems today to be one of the most effective. We must also consider that there are naturally numerous strains of bacteria resistant to antibiotics, particularly in the immediate environment of humans, because we exert strong selection pressure on them, consciously or otherwise.

4.1.2. Behaviour of the transgene in its new environment

When we introduce a DNA fragment carrying a gene with a role and function that are perfectly known, we cannot predict its behaviour in its new environment. It is in fact known that genomes are changing structures, malleable and adaptable, and that the same genes are expressed differently as a function of the genomes in which they are inserted. This behaviour could vary with the point of insertion and there could be interference with other genes: amplification, co-suppression, or disruption of a functional gene. These accidents could modify the behaviour of the transgenic plant, which could in some cases produce or secrete toxic substances. We are far from being able to predict the allergenic effects of products coming from GMOs. There have been some incidents in the United States with tryptophane produced from reprogrammed bacteria and in South America with a soybean transformed with introduction of a gene from *Nux vomica*.

We can also investigate the behaviour of the transgenic plant in open concurrence with its non-transformed congeners. Will it be quickly eliminated or will it on the contrary disturb the control population by developing new, unpredictable properties? Only subsequent experience will give us an answer to this question.

In this context, nothing can replace the value of isolation and multiplicity of tests, even though the advantage of time gained from genetic engineering will largely be cancelled out by the cumbersome, difficult, and long-term nature of the controls.

4.1.3. Flow and dissemination of transgenes

Among the reservations and uncertainties expressed is the dispersal of transgenes in the plant world. It is known that some plants of agronomic interest are relatively isolated in terms of reproduction, such as maize in Europe. The same is not true of other plants, which are known for the capacity to hybridize with plants growing in the wild or with adventitious populations. It could be feared from this that the gene introduced in a well-defined variety might naturally become incorporated in the genetic material of other species. This would be more serious in cases where the expression of the transgene must contribute to the distinction of species living near each other, as for example with the use of a selective weed killer. This fear is today more justified in that it has recently been shown that this scenario could be applied to rape, which has crossed spontaneously with wild radish (*Raphanus raphanistrum*) and mustard (*Sinapis arvensis*). Let us suppose a rape plant is transformed by a gene for resistance to a weed killer (glufosinate or glyphosate). Not only could the rape itself become a weed that is hard to eliminate from fields planted with other crops, it could also transmit its resistance to other Brassicaceae, which will then invade fields planted with cereal crops and resist all herbicides. Taking into account the considerable distance travelled by a pollen grain, whether by wind, insects, or birds, geographic isolation of the experimental field does not seem to be a sufficient guarantee and the threat foreseen could become a virtual agronomic catastrophe. This is the essential reason why transgenic rape crops have been placed under a moratorium since 1998 in Western Europe and are still under that moratorium, whereas for beet, which was restricted in the same way, the moratorium has now been lifted.

This is a serious problem that detractors of GMOs never fail to advance. It requires not only the exercise of great vigilance, but also new studies on the possible flows of genes across the plant community.

4.1.4. Appearance of resistance

Sexually isolated species could appear to be beyond the dispersal of transgenes. We have cited the case of maize, which does not hybridize with any other plant in Europe. The threat of hybridization does exist in other continents, as in the Americas, where there are still related wild species. However, it is in Europe that research programmes have been developed to eradicate a serious threat present in the form of a stem-boring insect, pyralid, which causes great damage and significant yield losses. We have seen that the larva of this insect can be controlled by chemical insecticides, but also by biological means, especially the use of *Bacillus thuringiensis*, which secretes a toxin that alters the cells of the intestinal epithelium. The larva, incapable of feeding, dies before moulting. Once the gene coding for the synthesis of the toxin was identified, researchers thought of transferring and inserting the gene for the toxin into the genetic patrimony of maize. Earlier assays on tobacco made it possible to validate this strategy and some advantages could be foreseen from the technique: specificity of the target, economy of inputs, protection of the water table. Pyralid-resistant cotton (Calgene) and maize (Novartis) were created, tested, multiplied, and developed to the point that they may now represent more than 50% of the crops in some regions. Encouraging results were recorded but phenomena of resistance appeared in some lines of insects. From that point, some anxiety arose, which was relayed and amplified by opponents of transgenic plants. New strategies had to be developed, such as the maintenance of parcels covered by non-transformed plants, parasitized by pyralid larvae, so that these insect populations would exert selection pressure on the resistant populations that appeared. Another controversy was raised about the effects of these insecticidal plants on populations of insect pollinators such as bees or protective insects such as ladybugs. A period of confusion followed, as is obvious from the position of French political authorities, which at first approved the import of transgenic maize while prohibiting its cultivation and subsequently allowed cultivation on a temporary basis. The same confusion prevails in the labelling rules imposed on GMOs and the products of GMOs, but not on products that are derived or transformed. These inconsistencies and hesitations have been exploited by detractors of these technologies.

4.1.5. Ownership of transformed plants

The debate on transgenic plants is also animated by a question that concerns law rather than biology: to whom do transgenic plants belong? The registering of a patent is a logical protection in the sector of conventional technologies, but it could raise controversy when it comes to a living organism. Can one become a proprietor of genes simply by registering a patent? These are the terms of a debate that has periodically stirred the agricultural sector, where patents have so far involved only farm machinery and the ownership of a living organism is only a matter for listing in the official catalogue. This type of approach has the merit of recognizing innovation without depriving other partners of the right to use the variety deposited for a new improvement programme. Progress becomes accessible to all. On the other hand, patents "privatize" the transformed plants, raise thorny problems of access to genetic resources, and encroach on control of production. These protections considerably increase the dependence of farmers on major corporations with advanced technologies, as well as the dependence of developing countries on industrialized countries. They reinforce the position of major corporations, which have the power to decide technological choices, the partners that will benefit from research, and ultimately the risks that society will run.

Awareness of these constraints and stakes has largely contributed to the setting up of legislative mechanisms designed to protect the right of citizens to choose the material conditions of their lives.

4.2. EVALUATION OF ADVANTAGES

The first section of this chapter may create the impression that there are only opponents to these new technologies. There are, however, scientists who do not wish to shut out any source of knowledge, and then there are supporters who see biotechnologies only in terms of quick returns. Many comparisons have been drawn between the development of biotechnologies and the development of nuclear energy during the 20th century, and we can find identical and equally inconsistent positions in both cases. There are resolute opponents of nuclear energy who would not tolerate daily load shedding, just as there are unconditional opponents of pesticide use who will only consume perfectly healthy and graded fruit. The supporters of the strictest moratoria must not forget that a pure and simple prohibition of biotechnology may send the corresponding research underground, with products that are difficult to identify and control.

4.2.1. A remarkable tool of knowledge

Plant biotechnologies and genetic engineering in particular undeniably play and will play a leading role in the *understanding and analysis of the genome* and of the critical function the genome serves in the growth and development of plants. Through these techniques, a certain number of genes of development have been identified. We can hope that such efforts will contribute to our understanding of secondary metabolism at the origin of molecules that can be exploited in physiological and pharmacodynamic terms. The mechanisms of response and resistance to stress will also be greatly clarified by it. Advances in knowledge are and must remain the primary reason for the research effort in this field. Nevertheless, we cannot refuse to exploit technically the knowledge that may fulfil essential demands for well-being and better living.

4.2.2. Productive and environment-friendly agriculture

Modern agriculture is efficient only at a high input cost and is today a major, measurable cause of environmental pollution. Moreover, despite its performance, it still has not met the needs of an ever-growing humanity. We cannot afford, therefore, to neglect any avenue that research may open up.

Among the favourable returns of biotechnologies, supporters cite resistance to herbicides as a technology well adapted to management of the biological environment of the plant by reducing the concurrence of weeds. They also consider the favourable impact on the environment of plants that require no insecticide treatments because they are genetically protected. They emphasize the advantage of being able to create plants particularly adapted to a specific industrial or economic activity—quality of starch, oils, reserve proteins—and consequently the considerable opening up of agricultural products to industrial purposes. The manufacture of drugs by plants also opens up vast perspectives for diversification of agriculture. Not only are the drugs products of high added value but there is also the consideration generally associated with anything that could improve human health. The most optimistic scientists hope to be able to introduce into the agronomic sector plants that are presently excluded from it because they produce nothing useful or produce something in a very small quantity. Biotechnologies have the power to activate genes. All the advocates of biotechnology ultimately agree that plants that have been genetically manipulated are infinitely more and better controlled than those from crosses and selection, and that food security has never been so great or so valued.

We can reasonably expect, therefore, a development of the biotechnological approach in agronomic research with the triple objectives of a more productive, less costly, and more environment-friendly agriculture. Plants that are cultivated on a large scale will be the first to benefit from these advances but all sectors of agriculture, horticulture, and forestry will benefit equally rapidly.

4.2.3. More focused industrial uses

Industrial uses of agriculture are sure to increase rapidly in the next few years. These advances require a close fit between the quality and availability of products and the needs and demands of industry. Agriculture must also adapt itself to uses of the industrial world, which requires *large quantities* and a *regular supply* but looks for the *lowest price*, whereas in the food sector the *quality* is often more important than the search for the lowest cost, even if the price begins to be more critical during periods of economic stress. In this search for suitability, a certain number of public and private organizations have been created, often with regional aims or competences, primarily to bring together partners of the two sectors in order to identify the needs of one sector and propose solutions to the other to allow them to meet those needs appropriately. The regional centres for innovation and technology transfer in numerous countries are examples of such structures, as are development organizations that have flourished, especially in some universities and within some official research bodies.

What has evolved on a regional scale and corresponds to partnerships involving small-scale producers has also developed on the national and continental scale. Today, we find agreements for very close collaboration that can go as far as buyouts between partners of the industrial sector (heavy chemistry, energy) and the growers. It is obviously very tempting to create by transgenesis a plant resistant to a herbicide and market it mainly to promote the sale of that herbicide. This concentration of two wings of a problem in a single hand confers a decisive competitive advantage that raises major concerns, especially when we observe that one-third of field assays of transgenic plants involve herbicide resistance. Another third involve resistance to pests and we find that the phytosanitary chemical industry (accounting for more than 30,000 millions Euros in the world) manages most studies on plant biotechnologies. The agrofood industry that agriculture has largely contributed to creating represents the rest and gives orders to the latter. However, since it is a sector that can be influenced by the opinion of the consumer, the development of plant biotechnologies is closely linked to their transparency and consumers' perceptions about them.

These favourable perspectives of plant biotechnologies in agriculture must not obscure an agonizing reality: despite the dazzling progress of agronomic science and technology and the increase in productivity, there is no generation of employment in agriculture and the question arises whether these technologies do not contribute sometimes to the collapse of certain sectors of agronomy that demand qualified workers. Moreover, do we take into account the number of seasonal workers who have lost their jobs by the introduction of a gene for male sterility in a plant of economic importance?

4.3. REGULATORY MECHANISMS

It is the balancing of decisive advantages against foreseeable risks that allows each of us to forge an opinion and defend the position that appears best adapted to the future of humanity. However, as the advantages and disadvantages of biotechnologies are not measurable values, the positions are highly varied and have very quickly necessitated the setting up of laws to regulate experimentation and especially the impact of experiments on daily life. These laws, national at first, have become increasingly streamlined across countries.

For example, in France the control mechanism is represented by two commissions: the Genetic Engineering Commission or CGG under the supervision of the Ministry for Higher Education, Research and Technology and the Biomolecular Engineering Commission (CGB), which depends on the Ministries for Agriculture and Environment. The CGG chiefly regulates research laboratories through approvals and the CGB manages applications for field assays and commercial production. All laboratories wishing to undertake studies using or designed to create GMOs must present an application for approval from the CGG and give guarantees on the confinement of the organisms. All studies requiring field assays must obtain the authorization of the CGB, which will strictly monitor adherence to cultivation conditions designed to prevent the dissemination of GMOs and their transgenes as well as monitor information to the public through notices.

Each profile includes a section translated into English for the benefit of other European countries that rule on the validity of assays and products. Downstream of these structures is a validation of the profile for all the EEC countries, presently 15 members but to include 25 members in June 2004. There are no similar structures on the global level but commercial exchanges about products derived from biotechnologies enter within the framework and cannot be in contravention of measures for free trade exchanges determined by the World Trade Organization (WTO).

These organizations, some of which have operated for twenty years, provide valuable information on the direction of research in the field of biotechnologies.

Nevertheless, it must be noted that the existence of legislation adapted to biotechnological products supposes the existence of two totally separate networks and the maintenance of that separation presently is one of the greatest difficulties to be resolved. Transgenic products are known to be present in traditional products. The term "GMO contamination" is often used but has a double meaning because a transgene is also likely to be diluted within a host population. For example, a massive introgression of transgenic DNA was observed in the 2001 harvest of maize in Mexico. As a corollary, an improved variety of cotton cultivated in Africa lost its utility following a transfer of wild genes. At present it appears that the two independent networks exist when the exchanges between them do not exceed a particular threshold of about 1 to 2%.

FURTHER READING

Droit et Génie Génétique. 1995. Biofutur/Lavoisier, Cachan, France.

Reports of the Academie des Sciences, France (published by TEC et DOC, Paris):

  no. 32, *La Brevétabilité du Genome*, 1995.

  no. 33, *Biodiversité et Environnement*, 1995.

  no. 36, *La Thérapie Génique*, 1996.

Report of the Academie des Sciences, Paris, *Les OGM*, Dec. 13th, 2002.

The journal *Biofutur*, particularly no. 206 (12/2000), no. 215 (11/2001), no. 218 (1/2002), no. 221 (4/2002), spec. no. 2 (10/2002) and no. 231 (3/2003).

The daily *Le Monde*, Sept. 23, 2001.

Sharma, K.K., Sharma, H.C., Seetharama, N., and Ortiz, R. 2002. Development and deployment of transgenic plants: biosafety considerations. *In Vitro Plant*, 39, 106-115.

The Future of Biotechnology and Genetic Engineering in Plants

A great deal has already been written on plant biotechnology but opinions often differ. It may be that the lack of distance from a science that is still very recent and multidirectional makes it difficult to draw up medium- or long-term projections and the sensitivity of each individual to these technologies has much more influence than objective reasoning.

In terms of fundamental research, we must expect a rapid development of biotechnologies. The genomic approach to a question or problem is now inevitable and clearly traditional disciplines such as biochemistry, cellular biology, physiology, and even systematics rely on the incomparable tool of genome analysis. Under the pressure of events, hard core objectors are every day smaller in number as they become aware of the power of this approach and the time it can save. This is why biotechnology must be part of any programme of biology. It is also true that this tool is a *complement* and will never replace other approaches or offer a definitive response or miraculous solution to all problems.

We can also anticipate a growth in applications of these technologies, but that will be closely linked to the capacity for communication and honesty of the agents and partners involved. Caution is the rule because any identifiable or predictable risk is unacceptable. The research must be totally transparent and the controls must never be subjected to economic pressures. Public opinion must always be taken into consideration, respected completely, and never overruled. Although public opinion will accept the taking of risks for responses to serious medical problems, there will be reservations when biotechnology affects what we eat, while organic products are popular and conventional agriculture improves day by day.

In view of the rapidity with which plant biotechnologies are being developed and applied, we do not have to be prophetic to affirm that it is not a simple mode without consequences and that it will have an increasingly prominent presence in our consciousness as well as our daily lives.

Exercises

1. Draw up a detailed plan for a small laboratory of about 16 sq m designed for *in vitro* plant culture, carefully indicating the material needed and how you plan to procure it. Also draw up a projected budget for setting up the laboratory.

2. Do the terms "seeds, grains, and plants" have the same meaning for a farmer, an agronomist, a trader, and a client?

3. An agro-food enterprise wishes to procure carotene and asks you to develop a supply plan. Specify the following:
a) Why does this enterprise need the product?
b) What will be your approach and what are the criteria that will guide you?

4. A lab assistant has for some days used samples of dermatological products sold by a famous cosmetologist. One morning, she is horrified to find that her face is slightly swollen and covered with small bumps, some of which have purulent cracks.
a) Does she have cause for alarm?
b) What type of simple analyses can she carry out in her laboratory to understand what has happened?
c) What measures must she take immediately?

5. An experiment in genetic engineering was carried out using the technique of leaf discs in co-culture with agrobacteria. The sequence of the transferred gene is known. This gene was introduced in a binary plasmid whose marker gene is that of neomycin phosphotransferase of type II derived from transposon Tn5. There are petri dishes with control and transformed cultures.

The indications marked on the petri dishes have disappeared because the ink used was not photostable.

What are the observations or tests that the unlucky researcher must make to re-establish the initial protocol?

6. The transformation of an *Agrobacterium* carrying nuclear resistance to streptomycin and rifampicin by a conjugant plasmid containing the gene for kanamycin resistance present in a colony of *E. coli* DH 5 α necessitates the presence of a third helper bacterium. This helper bacterium carries a plasmidic resistance to ampicillin, synthesizes a mobilization protein, and thus allows us to obtain a so-called triparental conjugation and the transfer of a plasmid carrying kanamycin resistance to the *Agrobacterium.*

a) What antibiotics must be used to supplement the culture mediums needed for the growth of the three types of bacteria?

b) In which cases is the presence of the antibiotic(s) indispensable?

c) On what medium are the different bacteria cultured after conjugation in order to recover only the *Agrobacterium* containing the plasmid that carries the kanamycin resistance gene?

7. Professor Nimbus' technical assistant is as absent minded as her director and somewhat careless. She was in charge of the greenhouse cultivation of tobacco plants (Solanaceae; autogamous) and lucerne plants (Fabaceae; allogamous). There were pots of control plants and pots containing transgenic plants. She left on holiday just as the plants flowered and returned eight weeks later to discover a virtual virgin forest with impressive quantities of aphids and other parasitic insects. She learned that during her absence her director had been kidnapped and later released through his colleagues had paid a huge ransom for him to be kept prisoner. Since his return, he has been living ostentatiously.

She collects as many seeds as possible but is worried about the plants that produced those seeds.

a) Does she have reason to worry?

b) What are the precautions she should have taken before going on holiday?

c) What are the characteristics of the germination tests she must carry out?

d) What explanations could she give for the sudden wealth of her director?

e) In what season should she take her holiday next time?

8. You are surprised to receive an invitation from the Minister for Agriculture to meet him the next morning at his office. He wishes to consult you on possible ways to reduce the deficit due to massive imports of plant proteins for animal feed.

You feverishly prepare for your meeting.

a) How will you get to his office at the stated time?

b) What documents will you carry with you?

c) What phrases will you use to direct the discussion?

d) The minister asks you for three suggestions: What do you propose to him? What arguments do you rely on?

e) What do you intend to hold back from such a discussion? In what state of mind do you imagine you will return from the minister's office?

f) You tell your plant biotechnology professor about your visit. He does not seem surprised about this meeting. Why?

9. On a small metre gauge line in central France, a station master is constantly fighting with the guard, whose guardhouse is located at least 400 m west of the station. This conflict is kept alive by the remarks of travellers, who never fail to compare the flowerbeds that the two protagonists actively tend in an open, constant rivalry that most often ends in favour of the guard.

The station master particularly accuses the guard of casting spells on his flowers and considers him responsible for a long list of anomalies:

- his dahlias and chrysanthemums have unexpected colours;
- his potatoes look like small tomatoes;
- his primroses are sterile;
- he cannot get rid of the wild radish and wild mustard plants that invade his garden in spite of his use of selective herbicides;
- none of the flowers he has planted around his walnut tree have germinated.

The guard consults you. What are the arguments that you can suggest to help him refute the station master's accusations, keeping in mind that the latter is also devoted to beekeeping?

10. The dean of a small provincial university, known for the reputation of its teachers and researchers, is passionate about botany and has an unlimited admiration for his colleagues who specialize in plant physiology. But when he creates a vegetable garden near his office he consults a professor recently appointed to a post of biochemistry, to the great surprise of the plant physiologists, whose abilities could have helped him avoid a horticultural catastrophe. Some months later, and after he has invested considerable effort, the result is far from what he wanted. He then consults you to attempt to discreetly remedy the various problems encountered.

You quickly draw up an evaluation and observe in particular the following:

- three out of four radish plants that germinated are affected by chlorosis;
- the leek, the only survivor of the flower bed, harbours a colony of nematodes;
- the three cauliflowers are relatively well developed but do not flower;
- one of the two artichokes is in full bloom;
- the only beet is also in full bloom;
- the aubergine plants have fruits shaped like tomatoes;
- the lucernes are magnificent;

- the carrots are all of the same size but lanky;
- the peas are very well grown but all mature at the same time and are particularly hard;
- the endive was green and highly acidic, and when the dean put it in a salad with potatoes that had been left for some time on the ground so that they would be washed by the rain, he got sick from eating the salad.

After having found possible scientific explanations for each of these problems, what simple solutions can you propose to help him salvage part of his reputation, if not his investment?

POSSIBLE ANSWERS

1. At least five items are essential in a laboratory for *in vitro* culture:
- an office with a computer, modem, telephone, fax, internet connection, and printer;
- a local technique for sterilization with *at least two access points* and equipped with an autoclave (controlled by official services) and pasteurizer;
- a sterile chamber (air filtered) equipped with a horizontal laminar flow hood for manipulations in aseptic conditions, supplied with water and gas;
- a culture room with controlled temperature, illumination, and humidity equipped with shelves for the culture of plants;
- a laboratory for the preparation of mediums and all the manipulations in non-aseptic conditions with assay balance, precision balance, a cabinet for chemicals, distilled water and ultra-pure water, centrifuge, filtration systems, etc.

The proposed budget can be established by consulting and comparing the tariffs proposed by the many companies that supply laboratory equipment and products, some of which specialize in *in vitro* culture. The equipment of such a laboratory can be estimated as follows. Amounts are given in Euros.

building	90,000
office	2,800
technical area	15,000
sterile room	4,000
culture chamber	23,000
laboratory	15,000
operations	12,500/year

2. A wheat or rye grain is a caryopsis, i.e., a seed with inner envelopes of fruit for an agronomist and a seed for the farmer. The potato plant is a rhizome, i.e., a succulent underground stem with plagiotropic growth, but the agronomist can create a new variety of potato from a seed or fusion of protoplasts. As for the market, some seeds are sold in bags, others in boxes, and still others in nets.

3. Carotene is a pigment of plant origin (polyisoprene) that accumulates in the chloroplasts. It is also synthesized by some yeasts. It constitutes a precursor of vitamin A synthesis and is not toxic. It is a permitted food colour and many products derived from the agrofood industry contain carotene.

Carotene can be extracted from the carrot root or aerial parts of other plants. It can also be extracted from microorganisms cultured *in vitro*. It is foreseeable that transgenic plants may be created that overexpress the gene for phytoene synthase. The strategy you propose will depend on the quantities required. It essentially takes into account the retail price, regularity of supply and quality of the product, storage systems, management, and eventual value addition of by-products or wastes.

4. The laboratory assistant will obviously worry and will certainly at first link the use of the cosmetics with the eruptions on her skin.

Nevertheless, she cannot rule out the fact that there may be no link (e.g., there could be hepatic disorders, contagious bacterial or viral infection).

Despite this reservation, she is right to ask whether the cosmetic product may contain an allergen. The answer cannot be found from a simple laboratory analysis. She may also suspect that the product contains DNA, as cosmetic products often seem to, and her worry may escalate to a fear that some of the cells of the generative layer of her skin may have been genetically transformed by this exogenous DNA. She may, however, be reassured by the fact that this DNA has undergone numerous processes of extraction that have probably denatured it and that the chances of its reaching and entering a cell colony are low. Moreover, the DNA needs to have retained the signals required to express and replicate itself if it is to insert itself in other cells.

At this point, it does not seem useful to look for an answer from a simple, discreet analysis. She decides to do the following:

- stop using the product;
- immediately consult a skin specialist and communicate her fears;
- ask the cosmetologist about the problems she has faced and find out the ingredients of the product so that she can take further steps.

5. The first approach is to test the material by culture on an antibiotic-enriched medium. The researcher must not, however, rule out the possibility that there may have been insertions of the marker gene without the target gene (recombination, breakage of DNA during the transfer) and that the use of a gene probe by Southern blot may be necessary. Such a decision can only be taken if the value of the transformed product justifies it. It is sometimes preferable to carry out a fresh transformation.

6. a) For the *Agrobacterium*, the medium must be supplemented with streptomycin and rifampicin, for the helper, with ampicillin, and for the conjugating plasmid, with kanamycin.

b) The antibiotics that supplement the mediums are not indispensable when the resistances are nuclear but are so if their origin is plasmidic (to avoid the loss of the plasmid by imposing a selection pressure).

c) *Agrobacterium* that has incorporated the kanamycin-resistant plasmid will be selected on a medium containing streptomycin, rifampicin, and kanamycin all together. Such bacteria will in fact be the only ones to be resistant to all three.

7. The technician is right to be worried. No one has monitored the cultures in her absence and crosses may have occurred, particularly in lucerne (allogams) and possibly even in tobacco, which is not a strict autogam. A serious laboratory technician would not take a holiday just when the plants are about to flower. She should have at least separated the plants by experimental category under self-fertilization bags or cages. She can still test the plants by carrying out germination tests on medium supplemented with antibiotic (kanamycin). Tests with specific probes and possible partial sequencing may confirm the conclusions of the germination tests.

The technician may suspect that her director may have planned to sell some of the genetically manipulated plants, which would explain why he did not object to her taking a holiday at this precise time of the year. At least he has probably favoured the departure of the technician so as to be able to fake a kidnapping and himself collect the ransom paid by his colleagues!

She decides to take her holiday in winter next time.

8. A request from a serving minister for a meeting with a student is rare enough for the student to take it seriously, especially since it would be a highlight on his resume.

He verifies that there are no train strikes exactly on the day of his meeting. Along with his umbrella, he carries his certificates and diplomas on the subject of agronomy and some journals of agricultural statistics, which can be obtained from the agricultural department, the regional directorate of agriculture, or farm cooperatives.

He begins by thanking the minister for this opportunity and assuring him of his keen interest in the mission. He nevertheless spells out the range of his competence and his very limited experience.

He does not actually advise the minister but draws his attention to the following: The decisions made by the minister cannot be free of humanitarian considerations with respect to the producers, the farmers. They must take into account consumers' opinions without rejecting summarily the scientific progress gained in plant improvement, including plant genetic engineering. They will be made in the context of new prospects not only for producers but also for young scientists who wish to work toward the qualitative improvement of products.

The student considers that this exchange, even if it does not have a concrete result, represents a valuable experience and a useful addition to his education.

The student's professor of plant biotechnology is not surprised perhaps because he feels the reputation of his lessons is already well know to the minister. At least he manages to conceal his bitterness at not having been consulted himself!

9. In central France, the wind blows mostly from the west and the distance between the station master's and guard's gardens does not make pollen transfer unlikely. You are convinced that there are scientific explanations for the facts presented by the station master.

You know that the recombination of characters in hybrids may intervene from the F_2 onward and give rise to plants that do not have the same genomic composition. You also know that some horticultural varieties that have undergone several hybrid crosses have characters that are not fixed and that the colour of flowers depends on the genes of the anthocyanin biosynthesis pathway.

Perhaps the station master planted only one of the two varieties of primroses (short style or long style) that must be crossed to result in fertile plants.

You explain that hybrids such as the potato-tomato or pomato could be obtained by somatic fusion of protoplasts.

You verify at the town council that there have been no assays of transgenic rape resistant to herbicides on the nearby commons. You explain to the guard that pollen from such plants could have fertilized and thus passed on resistance to the wild radish and mustard that are resistant to the station master's herbicide treatments. His bees could contribute to these interspecific crosses.

You know that walnut is teletoxic and produces large quantities of juglones, which are toxic to young seedlings.

There is no witchcraft involved!

10. The admiration the dean has for his plant physiologist colleagues is not enough for him to pick up knowledge about horticulture and gardening.

Chlorosis in radish can be explained by an insufficiency or non-availability of iron following its chelation.

There are antinematode treatments but they are preventive and must be applied very early. It is also possible to find lines that are less sensitive to this parasite.

Cauliflowers must be vernalized to reach flowering.

Artichokes have beautiful flowers but they must be consumed as buds.

Beet is a biannual plant. Its flowering mobilizes its sugar reserves

Aubergines and tomatoes belong to the same family and interspecific crosses cannot be ruled out. Protoplast fusions have also given rise to somatic hybrids and cybrids.

Lucerne is no doubt planted as a rotation crop for the following year. Lucerne requires no fertilizers because of its symbiosis with nitrogen-fixing bacteria.

Carrots may come from cloned artificial seed (homogeneity) but the planting density is not favourable to their growth. They are also dependent on nitrogen inputs.

Undoubtedly he has planted an autogamous forage variety of peas.

Endive is tender only when shoots are grown in the absence of light. Potatoes exposed to light develop "solanin", a toxic alkaloid.

You may advise the dean in future to confide his garden to a group of students who have some competence in agronomy.

Glossary

Acetyl CoA
Small molecule that contains an acetyl group linked to coenzyme A by an easily hydrolysed bond. This molecule constitutes a virtual metabolic framework within the cell and is found at the base of the vast majority of metabolic pathways: amino acids, shikimate, isoprenoids, etc.

Agrobacterium
Agrobacterium is a gram-negative soil bacterium that can parasitize higher plants, particularly Dicotyledons, and transfer to them part of its own DNA from the large plasmids that contain that DNA. The most commonly known is *Agrobacterium tumefaciens*, responsible for crown gall disease. It is widely used in plant genetic engineering because part of the DNA transferred from its Ti plasmid can be replaced with segments of DNA containing genes to be transferred.

Allele
One form of a gene. In diploid individuals, genes are present in the form of two alleles occupying the same locus on homologous chromosomes.

Antigen
Molecule that can be identified highly specifically by the immune system of an organism. For example, an antigenic reaction in plants is the pollen-stigma reaction during pollination preceding fertilization.

Apical
Describing cells that are located at the tip of the meristem.

Arabidopsis thaliana
Small plant of the family Brassicaceae, widely used in molecular biology because of the small size of its genome ($2n = 10$), its highly prolific reproduction, and its vegetative cycle, which can be reduced to 2 or 3 months (4 to 5 generations a year).

Antisense RNA
Complementary RNA of a transcript of a gene placed in the inverse direction of the native gene. It can thus hybridize with the RNA of this gene and prevent exploitation by polymerases and particularly its translation into a protein.

ATP synthase
Enzymatic complex in which the basal part is incorporated in the inner membrane of mitochondria or thylakoids of the chloroplast. This complex catalyses the formation of ATP from ADP and phosphate during respiration or photosynthesis.

Axenic (culture)
In vitro culture containing only a voluntarily introduced colony. Also called "pure culture" or, incorrectly, "aseptic culture".

Base pairs or bp
Two nucleotide bases of a DNA molecule paired by hydrogen bonds.

Capsid
Outer coat of a virus formed by self-assembly of repetitive protein elements called capsomeres, the whole being able to take well-defined geometric forms sometimes similar to those of crystals.

Cell fusion
Mode of union of two cells, often somatic, which combines their cytoplasm, membranes, organelles, and, in some cases, nuclear material.

Chaperon (molecule)
Protein molecule involved in migration or withdrawal of another molecule.

Clone
Population of identical molecules, cells, or organisms that descend by successive mitoses from a single molecule, cell, or organism without genetic modification.

Coenzyme A or CoA
Small molecule fixing on acyl groups (acetyl CoA).

Consensus sequence
Identical or partly identical DNA sequences sharing some genes that could have the same function or different functions.

Cosmid
Vector serving to clone and transfer segments of DNA and deriving from bacteriophagous viruses.

Cybrids
Organism deriving from the fusion of two somatic cells that have all the cytoplasm of two parent cells but generally the nuclear genome of only one.

Cytokinins
Family of molecules deriving from the mevalonate pathway and intervening in the cellular multiplication, growth, and development of plant organs.

Deletion
Loss of a DNA segment and the genes carried by that segment.

Diploid
Describing cells having all the pairs of homologous chromosomes and consequently both alleles of a single gene.

DNA library
A set of fragments of DNA molecules cloned in plasmids of bacteria and representing the entire genome of an individual (genome bank) or all the DNA copies of a genome, realized from messengers produced by cells at a stage in which the variety of messengers transcribed is greatest, for example in the embryo (bank of copy DNA).

DNA sequencing
Determination of the chain of nucleotides in the DNA molecule.

Domain
Free region of a long DNA molecule, located between two regions fixed at the lamina, within the nucleus. Also describes a region of a protein enjoying a certain physiological autonomy.

Duplication
Doubling of a DNA segment and the genes carried by that segment.

Endonuclease
Enzyme capable of hydrolysing the phosphodiester function of nucleic acids, single- or double-stranded DNA or RNA.

Enhancer
Regulatory sequences of a gene that can regulate the activity of a promoter.

Eukaryote
Cell of an evolved organism having a nucleus limited by an envelope and cytoplasmic organelles.

Exon
Segment of the coding sequence of a gene that is transcribed and retained after maturation of the transcript and the specificity of which is found in the protein. Opposite of intron, a region of the messenger that is eliminated during maturation.

Exonuclease
Enzyme capable of hydrolysing phosphodiester functions towards the ends of the DNA molecule.

G protein
Heterodimeric protein, linked to GTP and intervening in transmission of messages within the cell.

Genome
All the genetic information of a cell or organism. The maternal and paternal genomes form *the* genome of the embryo.

Genotype
Set of genes, i.e., potential genetic information of a cell or individual. Differentiated from phenotype, which is the realization of that potential.

Haploid
Describing a cell or organism that has only a single set of chromosomes. Applies to gametic cells, spores, and organisms that result from the development of these cells without fertilization.

Heterosis
Properties of hybrids that could exceed the qualities of the two parents.

Heterozygote
Diploid cell having genes in the form of different alleles at a particular locus.

Homeobox
Common sequence of DNA of around 180 bp present in some genes intervening in development and expressed particularly during the formation of certain organs.

Homozygote
Diploid cells having genes in the form of identical alleles at a particular locus.

Hybridization
Pairing of two complementary nucleotide sequences (DNA/DNA; DNA/RNA; RNA/RNA).

Insertion
Integration of a DNA segment into the DNA of a chromosome.

Intron
Non-coding region in a gene sequence. Introns are transcribed but excised from RNA during its maturation. Therefore, no manifestation of them is found in the protein.

Isoprenoid
Molecule belonging to a large family, intermediary in the sterol biosynthesis pathway, the basic element of which is a unit of five carbon atoms: the isoprene.

Kinase
Enzyme that transfers the phosphate group of ATP to a protein to phosphorylate it.

Ligase
Enzyme serving to suture or *ligate* two fragments of DNA molecules by the intermediary of a phosphodiester bond. This is a reaction requiring energy.

Liposome
Artificial structure composed of one or several lipidic bilayers that can contain various substances, particularly DNA.

Microinjection
Injection of molecules, particularly DNA fragments, into a cell using a micropipette mounted on a micromanipulator.

Mutation
Spontaneous or induced modification in the nucleotide sequence of a gene that can be transmitted to the descendants of the cell.

Northern blot
Technique of transfer of RNA on to a support membrane that can then be detected or identified using a specific probe.

Nuclear lamina
Fibrous layer carpeting the inner surface of the internal nuclear membrane and formed of proteins called "lamins".

Nucleosome
Basic structural element of chromatin constituting a "bead" that associates a DNA sequence to a "core" formed of an octamer of histones.

Oncogene
Gene conferring a cancerogenic character on the cell.

Oosphere
Name given to the female gamete in plants. Not to be confused with the ovule, the organ that contains the oosphere.

Operon
Linked genes having a common metabolic activity and often a common regulation.

Peptide signal
Sequence of a protein that determines its final location within the cell. It is particularly present to ensure the passage of a freshly synthesized protein from the cytosol to the inner cavity of the reticulum.

Plasmid
Small molecule of circular DNA independent of the principal genome of the bacterium, endowed with an autonomy of replication and often represented in several copies. Example: the Ti and Ri plasmid of agrobacteria. These plasmids are widely used for cloning DNA molecules.

Pleiotropic
Describing genes involved in multiple metabolic pathways and capable of having several phenotypic expressions.

Polymerase chain reaction or PCR
Technique to amplify DNA fragments *in vitro* following alternate cycles of temperature variations in the presence of a particularly heat-stable polymerase: the *taq* polymerase.

Polyploid
Describing a cell that has several times the base number of chromosomes. Many cultivated plants are polyploid.

Polysaccharides
Polymers of saccharides in the linear or branched form. Examples are starch and glycogen.

Prenylation of proteins
Covalent bonding of an isoprene and a protein.

Primer
Short DNA or RNA sequence that can be used, when paired with a matrix DNA, to synthesize nucleic acids by elongation in the 5'-3' direction. A double primer is used in some PCR techniques.

Probe
Fragment of a DNA or RNA molecule marked radioactively (by a hot probe), chemically (by an enzyme), or physically (e.g., by fluorescence) and used *in vitro* to label a complementary nucleotide sequence by hybridization.

Prokaryote
Organism lacking nuclear compartment, for example bacteria, cyanobacteria, and mycoplasts.

Promoter
Nucleotide sequence controlling the transcription of the reading frame of a gene. This is where the RNA polymerase fixes.

Proteoglycane
Protein linked to one or several chains of glycane (sugar).

Reading frame
Position in which the bases are distributed in triplets, each triplet coding for the fixation of an amino acid during translation.

Replication
Duplication of a complementary DNA molecule.

Repressor
Protein that is linked to a regulation region of DNA in order to block the process of transcription of the gene located immediately downstream.

Restriction enzyme
Enzyme belonging to a family of nucleases and capable of cleaving the DNA molecule in very precise places as a function of the succession of nucleotides present at the restriction site. The cuts may be straight or staggered. These enzymes play a fundamental role in the mastery of the genome and the development of genetic engineering.

Restriction map
Diagram of the genome or part of the genome of an organism on which are indicated the various sites of restriction by restriction endonucleases.

Retrovirus
Virus with RNA but requiring the synthesis of an intermediate DNA in order to replicate.

Reverse transcriptase
Enzyme extracted from retroviruses that synthesizes DNA from a matrix of RNA.

Ribonuclease
Enzyme that serves to hydrolyse phosphodiester bonds of RNA.

Somatic cell
Any cell of an organism other than the reproductive cells or their near parents.

Southern blot
Protocol based on the separation of DNA fragments by electrophoresis and their specific identification using radioactive probes.

TATAbox
Consensus sequence located in the promoter of genes in eukaryotic cells, generally between −170 and −80 bp before the beginning of the coding sequence. This sequence is characterized by the sequence TATAA and has a role in the initiation of transcription.

Thylakoid
Closed structural unit of the inner membrane system of the chloroplast.

Transcript
RNA derived from the transcription of DNA.

Transgenesis
Process of genetic transformation of a cell by introduction of foreign genes.

Transposon
DNA sequence that can move within a genome and that could have a different genetic reading or interpretation in its new environment.

Vector
Intermediary serving to transfer a gene from one cell to another. The vector may be a virus or a bacterial plasmid. A cloning vector is represented by the same elements but with the objective of carrying a gene that is to be cloned. A vector of expression is constructed so that the gene that it carries can be exploited by transcription enzymes.

Western blot
Process of separation of proteins followed by identification by a probe (antibodies, for example).

YAC
Artificial chromosome of yeast that behaves like a normal chromosome and follows the rule of cellular genetics.

Zygote or egg
Cell resulting from the fertilization of the female gamete by the male gamete and thus at least diploid.

Index